T0225037

SpringerBriefs in Applied Sciences and Technology

Computational Intelligence

Series Editor

Janusz Kacprzyk, Systems Research Institute, Polish Academy of Sciences, Warsaw, Poland

SpringerBriefs in Computational Intelligence are a series of slim high-quality publications encompassing the entire spectrum of Computational Intelligence. Featuring compact volumes of 50 to 125 pages (approximately 20,000-45,000 words), Briefs are shorter than a conventional book but longer than a journal article. Thus Briefs serve as timely, concise tools for students, researchers, and professionals.

More information about this subseries at http://www.springer.com/series/10618

Gitanjali Rahul Shinde ·
Prashant Shantaram Dhotre ·
Parikshit Narendra Mahalle ·
Nilanjan Dey

Internet of Things Integrated Augmented Reality

 Springer

Gitanjali Rahul Shinde
Department of Computer Engineering
STES's Smt. Kashibai Navale College
of Engineering
Pune, Maharashtra, India

Parikshit Narendra Mahalle
Department of Computer Engineering
STES's Smt. Kashibai Navale College
of Engineering
Pune, Maharashtra, India

Prashant Shantaram Dhotre
Department of Information Technology
JSPM's Rajarshi Shahu College
of Engineering (An Autonomous Institute)
Pune, Maharashtra, India

Nilanjan Dey
Department of Information Technology
Techno India College of Technology
Kolkata, West Bengal, India

ISSN 2191-530X   ISSN 2191-5318   (electronic)
SpringerBriefs in Applied Sciences and Technology
ISSN 2625-3704   ISSN 2625-3712   (electronic)
SpringerBriefs in Computational Intelligence
ISBN 978-981-15-6373-7   ISBN 978-981-15-6374-4   (eBook)
https://doi.org/10.1007/978-981-15-6374-4

This Springer imprint is published by the registered company Springer Nature Singapore Pte Ltd.
The registered company address is: 152 Beach Road, #21-01/04 Gateway East, Singapore 189721,
Singapore

Preface

Anything in this world can be achieved or overcome through the "Power of will"

—Bhagwad Gita

Internet of Things (IoT) integrated augmented reality (AR) book is intended to present concise and summarized contents that put forward advantages, technological requirements and perspective use cases of integrating AR. Due to rapid advancement in broadband technology, the Internet is widely available at a cheaper cost as compared to five years down the line. Due to this, the cost of connecting has been decreased drastically resulting in more number of devices connected to the Internet. This is realizing the notion of IoT where routine AR methods to view 3D objects are being transformed into interactive spaces. This IoT and AR integration can be done by associating context and related parameters with all operations. This integration leads to the need for investigating new algorithms for context sensing, classification of context based on the sources and for pervasive augmented reality. The book focuses on an overview of IoT, its emerging trends and technical building blocks which are required for developing real-world use case and applications. IoT application phases in terms of the proposed 8C model are key features presented in this book which will help IoT architect and developer to understand the fundamental phases in application development. This book also discusses a layered perspective of IoT architecture so that it will be easy for the reader to understand different functionalities and IoT components concerning each layer. In addition to this, the book also presents various indoor and outdoor IoT use cases along with the issues and challenges in the development of these use cases. Each use case is presented by considering actors, various services for the use case. All use cases are also elaborated with the possible scope for integrating AR to enhance the interactive nature and ease of use. Behavioral modeling using a sequence diagram of all the use cases is a key feature of this part of the book. This behavioral modeling approach will help readers to understand every use case in a more simple way.

The next part of the book presents an overview of context management in IoT and how it is important in AR applications. This part of the book also presents context-aware middleware and various context-aware architectures in detail along with their technical building blocks and technologies. Finally, this book concludes with how IoT is to be integrated with AR for enhancing the interactive nature of applications. The last part of the book presents and discusses an overview of AR as well as various design issues in IoT-AR integration. Emerging IoT-AR use cases are also presented in this book. IoT-AR general architecture is also presented so that reader can understand different subsystems in the architecture and how they are interrelated. Explanation of IoT-AR integration with small real-world scenarios is the key feature of this part of the book. Finally, this book concludes with the open research and practical issues in IoT-AR application development and future outlook.

The main characteristics of this book are:

- A concise and summarized description of all the topics.
- Use case and scenarios-based descriptions.
- The behavioral approach using sequence diagrams for explaining various functions in the application. This unique approach will certainly help readers for better understanding.
- Numerous examples, technical descriptions and real-world scenarios.
- Simple and easy language so that it can be useful to a wide range of stakeholders like a layman to educate users, villages to metros and national to global levels.

IoT and AR are now fundamental courses to all undergraduate courses in computer science, computer engineering, information technology as well as electronics and telecommunication engineering. Because of this, this book is useful to all undergraduate students of these courses for project development and product design in IoT and AR. This book is also useful to a wider range of researchers and design engineers who are concerned with exploring pervasive augmented reality and IoT further. Essentially, this book is most useful to all entrepreneurs who are interested to start their start-ups in the field of AR, IoT and related product development. The book is useful for undergraduates, postgraduates, industry, researchers and research scholars in ICT and we are sure that this book will be well received by all stakeholders.

Pune, India   Gitanjali Rahul Shinde
Pune, India   Prashant Shantaram Dhotre
Pune, India   Parikshit Narendra Mahalle
Kolkata, India   Nilanjan Dey

Acknowledgements

We would like to thank many people who encouraged and helped us in various ways throughout this book, namely our colleagues, friends and students. Special thanks to our family for their support and care.

We are thankful to Honourable Founder President of STES, Prof. M. N. Navale, Founder Secretary of STES, Dr. Mrs. S. M. Navale, Vice President (HR), Mr. Rohit M. Navale, Vice President (Admin), Ms. Rachana M. Navale, our Principal Dr. A. V. Deshpande, Vice Principal Dr. K. R. Borole and Dr. K. N. Honwadkar for their constant encouragement and inexplicable support.

We would like to thank Prof. Dr. T. J. Sawant Sir, Honorable Founder Secretary of JSPM Group of Institutes, for his continuous support and constructive guidelines. We are indebted to Dr. Ravi Joshi, Director of Planning and Development at JSPM for his guidance and motivation. We are thankful to Mr. Ravi Sawant and Prof. Sudhir Bhilare for their great support during this work. We are extremely thankful to Dr. Rakesh K. Jain, Principal of JSPM's Rajarshi Shahu College of Engineering (RSCOE—An Autonomous Institute), for inspiring us to write this book. We would also like to thank Dr. Avinash S. Devasthali, Vice Principal of RSCOE, for his constant support. A special thanks to Dr. Suresh N. Mali, Prof. Mahendra Salunke, and Dr. Nihar Ranjam for their continuous motivation. We are grateful to Dr. Prashant Kumbharkar, Dr. Ram Joshi, Prof. Santosh Borde and Dr. Deepak T. Mane for kind cooperation and encouragement. I would like to thank my friends Girish, Kiran, Manoj, Onkar and Sunil.

We are also very much thankful to all our department colleagues at SKNCOE, RSCOE, Techno India College of Technology for their continued support, help and for keeping us smiling all the time.

Last but not least, our acknowledgements would remain incomplete if we do not thank the team of Springer Nature who supported us throughout the development of this book. It has been a pleasure to work with the Springer Nature team and we extend our special thanks to the entire team involved in the publication of this book.

Gitanjali Rahul Shinde
Prashant Shantaram Dhotre
Parikshit Narendra Mahalle
Nilanjan Dey

Contents

About the Authors

Dr. Gitanjali Rahul Shinde has obtained her B.E. Computer Science & Engineering from Pune University, in 2006 with first class with distinction, India, and M.E. degree in Computer Engineering from Savitribai Phule Pune University (SPPU), Pune, India. She completed her Ph.D. in Computer Science and Engineering from Aalborg University, Copenhagen, Denmark. She has more than 11 years of teaching and research experience. She has published more than 40 research publications at national and international journals and conferences. She has edited books to his credit by De Gruyter Press and IGI Global. She has authored 5 books.

Dr. Prashant Shantaram Dhotre has completed his B.E. degree in Computer Science and Engineering from Shri Ramanand Teertha Marathwada University, Nanded, in 2004, and M.E. degree in Information Technology from Savitribai Phule Pune University, Pune, in 2010. He has been awarded with a doctorate degree in Computer Science and Engineering from Aalborg University, Denmark, in 2017. Currently, he is working as an Associate Professor in the Department of Information Technology at JSPM's Rajarshi Shahu College of Engineering, Tathawade, Pune, India. He has more than 16 years of teaching and research experience. He has 30 publications in international journals and conferences. One of his research papers was awarded as the "Best Paper" in an International Conference on Intelligent Computing and Communication ICICC – 2017 Springer Series. He has authored 4 books like

"Context-aware Pervasive Systems and Application" (Springer Nature Press), and "Theory of Computations" and "Machine Learning" (Technical Publication). A special book chapter was published in a book titled "Cyber Security and Privacy: Bridging the Gap", from River Publication in Europe. He is a Reviewer for the Springer Journal of Wireless Personal Communications and Reviewer for Elsevier Journal of Applied Computing and Informatics, IGI Global – International Journal of Ambient Computing and Intelligence (IJACI), etc.

Dr. Parikshit Narendra Mahalle obtained his B.E degree in Computer Science and Engineering from Sant Gadge Baba Amravati University, Amravati, India, and M.E. degree in Computer Engineering from Savitribai Phule Pune University, Pune, India. He completed his Ph.D. in Computer Science and Engineering specialization in Wireless Communication from Aalborg University, Aalborg, Denmark. He was Postdoc Researcher at CMI, Aalborg University, Copenhagen, Denmark. Currently, he is working as a Professor and Head in the Department of Computer Engineering at STES's Smt. Kashibai Navale College of Engineering, Pune, India. He has more than 20 years of teaching and research experience. He is serving as a Subject Expert in Computer Engineering, Research, and Recognition Committee at several universities like SPPU (Pune) and SGBU (Amravati). He is a senior member of IEEE, ACM member, life member of CSI, and life member of ISTE. Also, he is a member of IEEE transaction on Information Forensics and Security and IEEE Internet of Things Journal. He is a Reviewer for IGI Global – International Journal of Rough Sets and Data Analysis (IJRSDA) and Associate Editor for IGI Global – International Journal of Synthetic Emotions (IJSE) and Inderscience International Journal of Grid and Utility Computing (IJGUC).

Dr. Nilanjan Dey is an Assistant Professor in the Department of Information Technology at Techno International New Town (Formerly known as Techno India College of Technology), Kolkata, India. He is a Visiting Fellow of the University of Reading, UK. He is a Visiting Professor at Duy Tan University, Vietnam. He was an honorary Visiting Scientist at Global Biomedical Technologies Inc., CA, USA (2012–2015). He was awarded his Ph.D. from Jadavpur University in 2015. He is the Editor-in-Chief of the International Journal of Ambient Computing and Intelligence, IGI Global. He is the Series Co-Editor of Springer Tracts in Nature-Inspired Computing, Springer Nature, Series Co-Editor of Advances in Ubiquitous Sensing Applications for Healthcare, Elsevier, and Series Editor of Computational Intelligence in Engineering Problem Solving and Intelligent Signal processing and data analysis, CRC. He has authored/edited more than 75 books with Springer, Elsevier, Wiley, and CRC Press and published more than 300 peer-reviewed research papers. His main research interests include medical imaging, machine learning, computer-aided diagnosis, data mining, etc. He is the Indian Ambassador of the International Federation for Information Processing (IFIP) – Young ICT Group.

Chapter 1
Introduction

1.1 IoT Overview

Mark Weiser proposed notion of the ubiquitous or ever-present computing in 1998, which improve efficiency of human activities by keeping technology behind services. As per Marks Weiser, ubiquitous computing is:

> The third wave in computing, that is now beginning. First were mainframes, each shared by lots of people. Now we are in the personal computing era, person and machine staring uneasily at each other across the desktop. Next comes ubiquitous computing, or the age of calm technology, when technology recedes into the background of our lives.

Ubiquitous computing realized using various technologies, Internet of Things (IoT) is one of the prominent and potential technologies which empower ubiquitous access of devices irrespective of their location. In the digital era of the twenty-first century, everything is connected to the Internet and can be accessed in a ubiquitous manner to make the user's life easier and to execute work efficiently. In this context, efficiency means, "faster, lower user involvement, optimum use of resources, reduced energy requirement to name a few. Before going ahead with IoT, let us discuss a scenario without IoT.

> Rohini is a working girl living with her old parents. Besides her professional responsibilities, she also needs to look after her parents and take care of domestic duties. Every day she prepares a list of items she needs to buy from the grocery mall on her way to the office. Sometimes she forgets to prepare a list or go to the grocery mall. During office work, she is distracted by her worries about the health of her parents. In the case of a health emergency, she needs to call her family physician for medical help.

Since this is the story of not only Rohini, but of many who juggle multiple responsibilities of work and home, the IoT system with its advantages can come to their rescue. If Rohini lived in a smart home in a smart city, then the scenario would have been very different.

G. R. Shinde et al., *Internet of Things Integrated Augmented Reality*,
SpringerBriefs in Computational Intelligence,
https://doi.org/10.1007/978-981-15-6374-4_1

Rohini's home grocery items would be equipped with Radio Frequency Identification (RFID) tag. Information about these items would be stored on cloud/server with the help of Gateway. She could get notifications about which items she needs to buy using mobile App. Moreover, she will not get distracted during her office work by these notifications. The mobile app would check her location coordinates, and she will get notifications only when she is about to leave office. A grocery shop is also a smart shop. Hence, depending on her previous shopping, the grocery shop will give suggestions, provide directions to pick up the items which she needs to buy, and also inform about special offers.

If her parents are equipped with implanted health sensors, she can get their health status from anywhere through a mobile app and can get advice from a family physician. Health parameters are monitored by mobile App. If any parameter is found abnormal, then the mobile App will contact the physician. And depending on the severity of the abnormality the physician could suggest medicine or send for an ambulance to transport them to the hospital. In the first case, her parents could take prescribed medicine from "smart medicine box". With these facilities, she could work at the office with more concentration and enthusiasm.

The above scenario clearly demonstrates the need and benefits of IoT in day-to-day life. IoT is a revolutionary technology to connect day-to-day life things to the Internet. In the literature, various stakeholders have defined IoT in different ways as per their perspective and business domain. First definition of IoT is presented by EU in their CASAGRAS and is defined as:

IoT is global communication network infrastructure which links physical and virtual entities surrounding to us by taking advantage of massive data capture and communication functionalities. This global communication infrastructure is built on legacy Internet and network setups. The main functionalities offered will be device identification, sensor-internet sharing and search in order to build potential cooperative services and applications. These functionalities will have main characteristics like data capture, transfer of events, connectivity and interoperability [1].

International Telecommunication Union (ITU) has defined IoT as:

IoT represents global infrastructure for information community which will provide network and communication services by interconnecting physical and virtual entities surrounding to us using on available communication technologies [2].

The Internet Engineering Task Force (IETF) defined IoT as:

IoT will connect all available objects surrounding to us in order to provide seamless communication and services based on the context. The main components of IoT are RFIS tags, sensor nodes and smart devices which will enable the notion of ubiquitous computing [3].

According to Dorsemaine et al. [4], IoT is

Group of infrastructures interconnecting connected objects and allowing their management, data mining and the access to the data they generate.

As per Chen [5], IoT is defined as "machine-to-machine communication," and it merges physical word with information word through the Internet.

According to Botterman [6], IoT is

Things in IoT will have identities along with virtual personalities which will use intelligent interfaces to operate in smart spaces and interact within environment, user and social context.

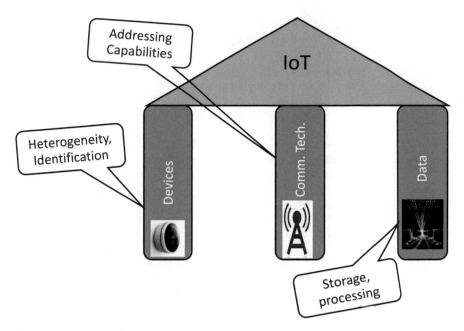

Fig. 1.1 Pillars of the IoT

Rayes and Salam [7] defines IoT as, *"It is the network of things, with clear element identification, embedded with software intelligence, sensors, and ubiquitous connectivity to the Internet."*

Despite such different definition of IoT, it consists of mainly three basic elements: things (devices), data (information) and communication technologies. These are the main pillar of the IoT which are presented in Fig. 1.1 with the need of respective pillar.

As stated in various definitions, things could consist of physical or virtual elements surrounding to us, i.e., sensor, actuators, objects with RFID tags, objects in indoor or outdoor use cases. Information is the parameters sensed by the IoT devices, and it may be an environmental parameter or user-health parameter and many more. These parameters must be converted to services to provide its benefit to the user. The third pillar of IoT is the communication technologies which connect devices to the user through services. Prominent communication technologies used in the IoT are Zigbee, near-filed communication (NFC), Bluetooth, Bluetooth Low Energy (BLE), RFID, cellular networks and IP wired/wireless network. IoT connects smart devices with these technologies and offers benefits to end users.

1. Things: Devices or objects used for collecting/sensing information. Devices are heterogeneous in size, computational capabilities, and energy source. IoT devices are smart devices with unique identification and sensing and communication capabilities.

2. Data: Data is the piece of binary or massive information collected from various IoT devices deployed and provided for further processing in order to decide context. On top of processed data, IoT services are defined to make users life better.

3. Communication Technologies: It is the backbone of communication in the IoT. This is the interface between the user and the device to get the benefit of IoT services.

To understand IoT, we need to define service built on data sensed by devices and communication technology. The concept is further explained by providing examples of the motivational scenario given in the previous paragraphs.

1. Devices: Device could be defined as any physical "thing" that has sensing and communicating capability. For example, smart phones, home grocery items with RFID tags and sensors; medicine boxes; smart trolley and the items equipped with RFID tags at grocery mall, implanted health sensors, i.e., blood glucose sensors, temperature sensors, blood oxygen sensors, ECG sensors, image sensors, motion sensors, inertial sensors, etc.

2. Services: The term services has various meanings according to the context. Here, services refers to the benefits that user gets from IoT devices. Information sensed by IoT devices is stored and processed at data servers/cloud. On processed information, services are built to provide benefits to the user. These services can be termed as "IoT Services."

    For example, if Rohini's parents are equipped with body sensors, the temperature sensor senses the body temperature and forwards this information to the server. At the server, this temperature value is converted into "body temperature service" to make it accessible to the physician and Rohini.

    Refrigerators are embedded with level sensors, RFID tags, and readers which send to the server the expiry date of item and quantity. And then, this information is converted into service. This service recommends Rohini to purchase an item that has crossed expiry date or is not in stock in sufficient quantity.

3. Communication Technology: It is a medium for data transmission from device to the server, server to the user, user to user and device to device. e.g., RFID, Bluetooth, near-field communication (NFC), Local Area Network (LAN), Wi-Fi, cellular networks and WiMAX and many more.

The layered perspective of the IoT is depicted in Fig. 1.2.

IoT has wide application domain, and it is almost important in every Internet application. The role of every IoT application is to provide IoT services to the user. Providing services to a user according to their needs and requirements is a major challenge of IoT. Concisely, we can say IoT connects people to devices to make their life easier and happier. IoT includes Internet of People (IoP) and Internet of Services (IoS).

The newly emerging concept IoP consists of a combination of IoT and people and is depicted in Fig. 1.3; it is nothing but people with gadget having Internet connectivity. It deals with how human and things can cooperate for shaping a smarter

Fig. 1.2 Layered perspective of the IoT

Fig. 1.3 Definition of Internet of People (IoP)

world. It is quickly spreading into the structure holding the system together giving a burst of new development to add to the facilitating development of cell phones, tablets and other traditional individual hardware and related systems and administrations. Numerous web-empowered peripherals and options are arriving that are worn, implanted in materials and items. This is on account of new materials and methods for making gadgets and more appropriate human interfaces. IoP wave is moving communication from machine to machine (M to M) to person to person (P to P).

IoP helps to give answers to many different questions, such as is cooperation of sensors, the intelligence of cloud, and humans will help to do things differently? How these smart things will change the perspective of human life? It is necessary to

Fig. 1.4 Definition of Internet of Services (IoS)

follow some specific strategy to make IoP beneficial. The strategy should be thought in view of the social and personalized behavior of people and things, and it should also consider proactive nature of heterogeneous devices and predictability of interactions [8].

The term IoS has different meaning according to context. Here, services are referred to benefits that user gets from IoT devices [9]. The Internet of Services referred to the collection of different services available on the Internet and provided through an IoT framework, definition of IoS depicted in Fig. 1.4.

The success of every IoT applications depends upon which service it provides? How innovative service it is? And how one application can provide maximum services? The new service development process comprises of service innovation, the design of service, and launch of the service. Service innovation is a customer-centric process. It is the way toward formulating another or enhanced service idea that fulfills the client's neglected needs. Service innovation should take care of uncovering customer needs and priority of needs. The service design makes the strategy based on inputs from service innovation. Service design alludes to all exercises required in actualizing this idea and offering it for sale to the public [5]. The future of IoT depends on the context management, innovation of new services and framework for adapting that services.

From the above discussion, we can define IoT as

> IoT is smart service oriented communication networks consisting of heterogeneous devices including resource constrained and resourceful and will be the mandatory part of future Internet. The main components of IoT are RFID objects, sensor nodes and smart phones. IoT connects everything surrounding to us and generates big data which needs to be posted on cloud and also this IoT – cloud convergence should be backward compatible with the legacy networks and communication protocol.

IoT majorly answers the question "how to collaborate People with IoT and Personalized Services." Here comes role of context management and augmented reality (AR), IoT with context and AR maximizes value associated with IoT services.

IoT connects devices to Internet while AR connects people to IoT devices by interacting with the physical environment. For example, *"IoT might tell me that a machine is going to have a problem and I can use that information to direct a worker where to go and what to do when they get there using AR"* [10]. Decision making is the major function of the IoT, and it is done by learning user preferences requirements based on user's context, history of usage and many more factors; hence, context management plays an important role in IoT and AR.

1.2 Emerging Trends in IoT

IoT and AR are in the evolution phase; many technologies are adding benefits to IoT; and evolution in IoT devices is also happening in faster speed. Nowadays, researchers are successful in many aspects like battery requirements are reduced in higher extent and the processing, sensing, communication and responding power of IoT devices are improved in great scale. Communication is the basic pillar of the IoT and AR; the communication technologies also adapted changes required for IoT and AR. Few such technologies are discussed in this section:

Zigbee: It is IEEE 802.15.4 standard, defines two layers as physical layer (PHY) and medium access control sublayer (MAC). ZIGBEE standard was defined on top of IEEE 802.15.4 used in wireless personal area network (WPAN) [11]. ZIGBEE system consists of three different types of devices such as router, end device and coordinator, depicted in the figure. Every ZIGBEE network must consists of at least one coordinator which is responsible for handling, storing, receiving and transmitting data. ZIGBEE Coordinator is used for formation of a network tree. ZIGBEE router acts as mediatory device used to pass the data to the coordinator device. End device is low power device with limited functionalities to communicate with the parent node.

Zigbee is an open, packet-based protocol designed for low data rate, low power, secure and reliable for wireless networks. Zigbee standard was specifically designed for monitoring and controlling personal area sensor networks (WPAN). Zigbee operates at 868 MHz, 902–928 MHz, and 2.4 GHz frequencies. The data rate for communication is 250 Kbps. The typical communication range is 30 M for indoor and 100 m line of site for outdoor application. Each ZIGBEE network can have 64 K number of nodes.

Zigbee device discovery is initiated by any Zigbee device by sending its own 64 bit address. After device joins Zigbee network, it receives 16 bit address PAN ID. After Joined network, device can send commands to other devices on the same network. There are three network topologies as star, cluster tree and mesh networks. The typical application of ZIGBEE is intelligent home automation system for home security purpose. The home automation system is used to control all the lighting, electrical appliances, automatic door opening, gas detection using ZIGBEE, GPS and sensor modules.

Wi-Fi: Wi-Fi is a technology based on IEEE 802.11 standard for wireless local area networking. WI-FI stands for wireless fidelity. WI-FI devices can connect to the Internet using access point or hotspot. WI-FI operates at ISM 2.4 GHz and 5 GHz frequencies. The data rate for communication is 54 Mbps. The typical communication range is 20 M for indoor and 100 M for Outdoor application. WI-FI technology uses TCP/IP stack for Internet connectivity. WI-FI is a cost-effective and most popular solution due to WI-FI alliances to run an interoperability programs on WI-FI devices. WI-FI is integrated in all laptops, tablets, TV and smart phone. WI-FI uses star topology where the center node is Internet gateway. WI-FI devices require large amount of power as compared to Zigbee and Bluetooth devices. IEEE 802.11 is a set of medium access control (MAC) and physical layer (PHY) specification for wireless local area network communication [12]. There are several 802.11 specifications as mentioned below,

- 802.11a: This is an extension to 802.11 pertains a wireless LANs which operate at 5 GHz band that goes as fast as 54 Mbps speed. It uses orthogonal frequency division multiplexing (OFDM).
- 802.11b: This is a high-rate WI-FI extension to 802.11 pertains a wireless LANs which operates at 2.4 GHz band that goes as fast as 11 Mbps speed. It uses direct sequence spread spectrum (DSSS).
- 802.11g: It provides 20 Mbps in 2.4 GHz frequency band

Wi-Fi device scans other devices to establish connection. Two types of scanning methods are used in Wi-Fi, passive and active scanning. In the passive scanning, device scans for probe request or beacon frame from another device/AP. In active scanning, device transmits probe request and waits for probe response from the AP. Probe request may be unicast/broadcast. In both types of scanning, active scanning is faster; however, it consumes more energy than passive scanning. Wi-Fi beacon frame is depicted in the figure.

Wi-Fi systems are half duplex where all stations are involved in transmitting and receiving the data on same radio channel. Radio channels cannot receive while transmitting the data; hence, it is impossible to detect data collision on a radio channel. To avoid this, mechanism was developed which is called as distributed control function (DCF). In DCF, WI-FI system will initiate transmission provided the channel is clear. All transmissions are acknowledged; if station does not receive acknowledge, then, it will retry to send the data.

Bluetooth, Bluetooth Low Power Energy (BLE) and Bluetooth 5 Bluetooth is the short-range wireless communication medium used to talk two devices that use radio frequency 2.4 GHz ISM band. Bluetooth operates in 2400–2483.5 MHz worldwide operation with 79 channels of 1 MHz each. The range of Bluetooth communication is 10 m. Bluetooth devices connection uses master/slave or client/server configuration. Bluetooth devices can send data in full duplex mode with 64 Kbps data rate [13].

When two Bluetooth devices want to communicate each other, they first need to pair each other. A piconet is a network of devices connected using Bluetooth. In piconet 7, devices can be connected to master station. An additional device can be attached in Parked or Hold state. Piconets are established dynamically when Bluetooth devices enter and leave radio proximity. In Bluetooth communication, one device acts as a master and other as a slave.

Bluetooth low energy was discovered to resolve the problem of high current consumption of the Bluetooth. Bluetooth low energy consumes approximately 15 mA which is very less compared to classical Bluetooth with a current consumption 30 mA. Bluetooth low energy reduces implementation cost, enhanced communication range and multiple vendor interoperability. Due to reduction in implementation cost and allowing BLE applications for numerous different venders simultaneously makes

BLE more reachable. BLE-enhanced range is useful when sensors are located where power and access are limited. BLE operated at a range of 250 m.

BLE is a subset of classic Bluetooth and not backward compatible with classic Bluetooth devices. There are two types of devices, single-mode and dual mode where dual mode devices operate as Bluetooth device and BLE device. When two BLE devise wants to communicate with each other, the first step is to initiate a connection request between BLE device and central device where BLE device can be a phone and central device can be the tablet. This process is called advertising process. This advertising process enable BLE device to be visible to the outside world. In advertising, a peripheral device sends advertising data to multiple central devices. The Generic Access Profile (GAP) enables controlling the connections and advertisement within BLE. There are two types of payload when advertising with GAP:

- Advertising data payload
- Scan response payload.

Scan response payloads are not mandatory and requested by central device. Scan response payloads allow to send additional information such as device name. Advertising interval can also be increased to reduce current consumption. After receiving an advertisement frame from BLE device, the central and peripheral device will create a dedicated connection. The advertising process is then finished, and the peripheral device can only communicate with one central device.

Further, recent advancements in Bluetooth technology is Bluetooth 5, which makes it a promising technology for the IoT. It is revised as per IoT requirements. It provides double data rate than BLE, i.e., 2 Mbps data rate. It also gives four times more transmission range than BLE by increasing transmission power level to + 20 dBm. Furthermore, Bluetooth 5 has eight times more advertisement capacity than BLE, which is based on features like secondary advertisement. This feature is more important in IoT and AR to broadcast various IoT services to the end users.

LoRaWAN: LoRaWAN is a protocol for low-power wide area network (LPWAN) [14]. It supports mobility, low powered and bidirectional communication; hence, it is suitable to use in the IoT network. It supports data rate of 0.3–50 Kbps. Using adaptive data rate scheme, LoRaWAN minimizes energy consumption of the end devices and improves the network capacity. It follows the star topology where the gateway is the relay node to transfer messages from end devices to the central server. It uses frequency band of 2.4 GHz and 5 GHz similar to the Wi-Fi. End devices are connected to the server by initiating join procedure. Join procedure consists of two MAC messages, join request and join accept. For secure IoT network, LoRaWAN consists of three different keys, i.e., unique network key (64 bit), unique application key (64 bit) and device specific key (128 bit).

Among above-mentioned technologies, BLE is suitable for applications that require low range, less energy consumption and low data rate communication. Wi-Fi can be used in the application where high communication range is required.

Table 1.1 Comparaison of communication technologies

Technology/parameters	Wi-Fi	Zigbee	BLE	LoRaWAN
Communication range (m)	50–100	10–100	10–70	2–5 km
Power consumption	High	Very low	Low	Low
Infrastructure requirements	Yes	Yes	No	Yes
Ready to use in smart phones	Yes	No	Yes	
Data rate	11-54 Mbps	250 Kbps	64 Kbps	0.3–50 Kbps

Zigbee can be used in the industrial application rather than in residential applications as Zigbee cannot readily use with smartphone Zigbee. The comparison of these communication technologies is shown in Table 1.1.

1.3 Technical Building Blocks of IoT

Most of the IoT devices are resource constrained in all sense. i.e., low processing power, low energy source, less computational capability and low range communication capability. Hence, for each IoT network, there is need of device which can work as gateway between resource-constrained network and high capability network, i.e., IP network. In IoT network, unstructured data is generated and that is in huge amount; on this data, services are built. IoT devices are resource constrained and due to security reasons, we cannot make these devices public. Hence, IoT data is stored on cloud and services are built on cloud. The devices, gateways, network infrastructure and cloud infrastructure are building blocks of the IoT network.

- Devices: This is the first layer in architecture which includes things like sensors, actuators which enable the communication and collection of information from the objects of underlined use case proactively. i.e.. without human intervention.
- Gateways: This acts as a second block or layer enabling the strong connectivity between the things and cloud infrastructure. This layer also provides the security and manageability abilities during the data exchange.
- Network infrastructure (NI): This block controls the data flow from first layer, i.e., device layer to cloud infrastructure which is described below. It also enables the security during the information flow between Routers, Aggregators, Gateways, Repeaters.
- Cloud infrastructure (CI): This is an application layer of the ecosystem involved in analytical, logical and advanced computing abilities. Virtualized servers (VS) and data storage units (DSU) are the main components of this block.

We can understand building blocks of the IoT system using 8C model; IoT follows the 8c model for seamless communication between user and IoT device based on above building blocks of the IoT, i.e., connectivity, communication, cooperation, convergence, context and content. Here, we propose 8C model of , i.e., connectivity, communication, cooperation, convergence, context and content presented in Fig. 1.5:

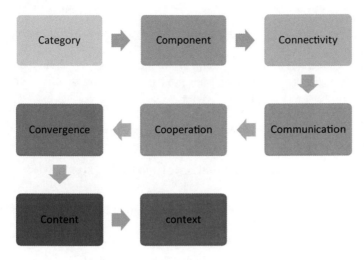

Fig. 1.5   IoT 8c model

- Category: Categorization of IoT applications is required before realization of application, as indoor applications have different requirements than outdoor applications. Selection of IoT devices, communication technologies, data storage and service distribution depend on application category.
- Component: The device classification plays important role in 8C model as based on the device capabilities roles can be assigned to the device. The device classification can be done based on various parameters, depending on the application requirements.
- Connectivity: In the IoT system to provide communication between IoT device and user, both must be connected to each other. Devices could be connected to the user using different communication technologies, i.e., Ethernet, Bluetooth, WLAN, Zigbee, etc. These differ in communication range, energy requirement, data rates and interface. Communication technologies used to connect the devices will depend on the type of devices. Connectivity needs unique identification to the IoT devices.
- Communication: In the IoT environment, different types of communications are needed, i.e., device to device, user to device and user to user. It may happen that IoT devices and user devices have different communication interface. A mechanism is needed to establish communication between heterogenous devices.
- Cooperation: Services are formed on the basis of information sensed by sensors. For more complex services, integration of more than one sensor is required. To achieve this, a close cooperation between different IoT devices is required.
- Convergence: Different types of IoT devices are connected to each other using various types of communication technologies that generate real-time, unstructured data. IoT services are defined on such heterogeneous data, which are

made available to the user for their betterment. For deployment of IoT application, convergence of these different technologies and data processing methods is required.

- Content: Enormous amount of data, i.e., "big data" is generated in the IoT applications in real time. Processing, managing and providing security to big data is required to form IoT services
- Context: To enhance the performance of IoT and AR system, the context information needs to consider. The mechanism for context fetching, context processing and managing is necessary in the IoT.

To realize the above scenario using 8c IoT model, IoT system needs to address the requirements listed below:

1. Devices must be organized systematically; group of IoT devices is formed based on some logical connection between them.
2. Among all devices, a device with strong capabilities should be selected as group head to collect data from group members and forward it to the server through GW.
3. The server should have a mechanism to process and manage data sensed by IoT devices. IoT services are defined on top of real-time data which is processed by the server.
4. For users to take advantage of IoT services provided by a group of devices, the advertisements should be in a format which is easily comprehended by the user. For personalized and context-aware service discovery, a mechanism to retrieve user preferences and context information is required.
5. The strong authentication and access control mechanism should be applied, and as the system is dealing with very sensitive data with poor security measures, IoT could result in disasters, e.g., if a hacker wrests control of the car and disables the car brakes, directs the thermostat to increase temperature levels to the highest or to uncomfortable levels, or manages to get control of wireless insulin pump and changes insulin dose [7]. David Jacobi has demonstrated the importance of security by hacking his own smart home [15].

1.4   IoT Layered Architecture

IoT follows the layered architecture; IoT has mainly four layers,. i.e., device and a sensing layer, communication layer and application and authorization layer [16] as depicted in Fig. 1.6.

Device and Sensing Layer: This layer mainly focuses on the organization of devices. Due to heterogeneity and constrained nature of IoT devices, it is important to organize them in a systematic manner. Device layer works on the question: How to connect IoT devices with the Internet? Not all IoT devices are capable of connecting directly to the Internet, and it is not ideally possible that each device has same communication interface, and therefore, here comes the role of a grouping of devices and gateway. Devices

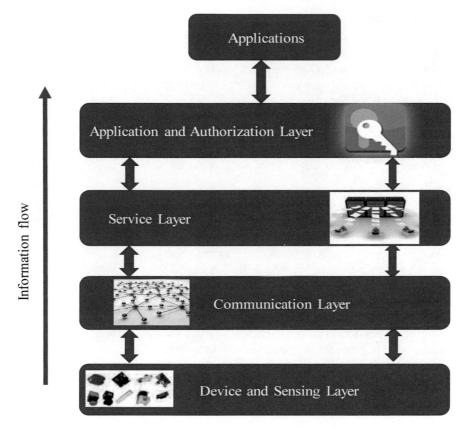

Fig. 1.6 IoT layers

with the low computational capability and heterogeneous communication interfaces are grouped together, and one gateway is allocated to each group. Gateway has different communication interfaces to connect to all devices of the group. Gateway mainly works in a scenario where communication protocol of devices is not same as the protocol used at the network layer. Device layer includes communication technologies such as NFC, RFID, Bluetooth, BLE, Zigbee, Z-wave, LoRaWAN, LAN, WLAN and cellular 3G, 4G network.

Primary Challenge of Layer: Organization and classification of devices, connectivity and heterogeneity.

Communication Layer: communication layer works as the interface between the user and IoT devices to get access to the IoT services. Communication must have the capacity to provide unique identification to the individual IoT device to connect it to the end users. Technologies used in the communication are IPV4/IPV6, 6LowPAN, IPv6 routing protocol (RPL), TCP/IP and UDP.

Primary Challenge of Layer: addressing and identification, low power communication, low memory routing protocols, non-glossy communication, mobility.

Service Layer: This layer stores and processes the information gathered from IoT devices. This information is in real time and unstructured in nature; hence, information processing is a very critical task. Service layer defines services using the processed information and makes users aware about IoT services. IoT services are different than other web services as these are dependent on the information sensed by IoT devices. Hence, service discovery and service enforcement is a difficult process. The technologies used for service layer are physicalweb, HyperCat, universal plug and play (UPnP), multicast Domain Name System (mDSN), DNS service discovery (DNS-SD), etc.

   Primary Challenge of Layer: store and process information, define IoT services.

Application and Authorization Layer: Providing IoT service to the user is the main function of application layer with the help of different IoT application protocols, i.e., Extensible Messaging and Presence Protocol (XMPP), Message Queuing Telemetry Transport (MQTT), MQTT for Sensor Networks (MQTT-SN), Advanced Message Queuing Protocol (AMQP), Constrained Application Protocol (CoAP), Data-Distribution Service for Real-Time Systems (DDS), etc. The main intention of IoT is to make users life simpler and comfortable by providing services, but not at the cost of hampering their security and privacy. To maintain user's security and privacy, it is necessary to verify identity and access rights of users before providing access of IoT services. This layer verifies the user's authentication and authorization. Access policies are set by service owner using different access control mechanism, i.e., role-based access control, attribute-based access control and so on.

   Primary Challenge: Provide IoT services to intended users by verifying their identity and access rights.

1.5   Issues and Challenges

IoT has many more technical and business challenges like standardization, connectivity issues, heterogeneity, energy demands, reliability, scalability, mobility, interoperability, waste disposal, storage, security and privacy [17–25]. From the above discussion of IoT system requirements, few challenges are mentioned below, and these are focused in this thesis.

1. **Mobility**: IoT system consists of many different types of devices. Every IoT device has some fundamental functionalities, e.g., sensing air humidity, current temperature, position, etc. Collaboration of more devices is required for complex functionalities. These devices could be static or mobile. To collaborate different IoT devices to provide complex functionalities, it is necessary to organize an IoT device in an efficient way. However, due to the mobility of IoT devices, it is very difficult and challenging to organize them.

2. **Heterogeneity**: IoT devices are heterogeneous in several respects. The IoT devices differ in size, computational capability, functionality, a communication

interface, energy source and many more parameters. To form IoT services with the help of such diverse devices is a major challenge for any IoT system.

3. **Scalability**: as mentioned in the introduction section, the number of IoT device deployment has increased tremendously and will continue to increase in the future. Most of IoT applications need a collection of a large number of static/mobile and heterogeneous IoT devices, and it is possible that number of devices in the application may change dynamically. To organize, connect and manage such a large number of devices, the IoT system needs to support scalability.

4. **Energy Efficiency**: In the IoT environment, most of the devices are battery powered. Devices need the energy to sense data, data transmission, data processing and for taking action depending on sensed data. IoT systems should be energy efficient to increase the life of the IoT devices.

5. **Security**: IoT systems are tremendously vulnerable to attacks for numerous reasons. First, most of the time devices are unattended, and consequently, it is easy to physically attack them. Second, most of the communications are wireless, which makes it extremely simple for an adversary to succeed. Lastly, most of the IoT objects are constricted by finite resources like battery, computational power and memory, and hence, complex security techniques cannot be implemented.

6. **Privacy and context management**: Service provider (SP) asks the user to accept policies with terms and conditions about sharing their personal data before providing access to the service. The user must accept policies to get the benefit of services; it is a "take it or leave it" scenario. Sharing personal data with the SP could possibly result in the invasion of the privacy rights of the user. As in the IoT environment, a large number of services are available. Hence, user data will be shared with more number of SP leading to possible infringement of privacy rights. Therefore, minimal personal information disclosure mechanisms are required to access services. Hence, for privacy-aware IoT applications, there is need of mechanism which can compute user's context and preferences based on minimal information shared by the user.

7. **Context management and knowledge retrieval**: IoT and AR is based on the context information; hence, processing of context information and retrieval of knowledge from are major challenges of the IoT and AR, as context information is heterogeneous in nature.

1.6 Conclusions

In summary, we can say that IoT is the service network which is converge of sensor nodes, RFID objects and smart devices. IoT connects objects around us (electronic, electrical, non-electrical) to provide seamless communication and contextual services provided by them. Communication technologies like Wi-Fi, Bluetooth, Zigbee, LoRaWAN are improved in great scale to be suited for IoT. Context information is important pillar to realize IoT and AR; hence, framework for context retrieval and processing is required. Various case studies of IoT and AR needed to discuss for understanding requirements of IoT and AR framework. In the next chapter, case studies of IoT and AR are discussed.

References

1. Project CASAGRAS (2009) CASAGRAS Final Report: RFID and the Inclusive Model for the Internet of Things
2. Overview of the Internet of Things Y.2060 (2012) In International Telecommunication Union (ITU), pp 1–18
3. Myoung L et al (2011) The Internet of Things concept and problem statement draft lee IoT problem statement status
4. Dorsemaine B, Gaulier JP, Wary JP, Kheir N, Urien P (2016) Internet of Things: a definition and taxonomy. In: Proceedings—NGMAST 2015: the 9th international conference on next generation mobile applications, services and technologies, pp 72–77
5. Chen Y-K (2012) Challenges and opportunities of Internet of Things. In: 17th Asia and South Pacific design automation conference, pp 383–388
6. Botterman M (2009) Internet of Things in 2020: roadmap for the future
7. Rayes A, Salam S (2017) Internet of Things (IoT) overview, in Internet of Things from hype to reality. Springer International Publishing, Cham, pp 1–34
8. Miranda J et al (2015) From the Internet of Things to the Internet of People. IEEE Internet Comput 19(2):40–47
9. Soriano J et al (2013) Internet of services. In the convergence of Telecom and Internet on evolution of telecommunication services, vol 7768, pp 283–325
10. McKendrick J (2019) Available online: https://www.rtinsights.com/augmented-reality-is-iot-for-people/
11. Alliance, Zigbee, Zigbee Specification (2008) Available online: https://people.ece.cornell.edu/land/courses/ece4760/FinalProjects/s2011/kjb79_ajm232/pmeter/ZigBeeSpecification.pdf
12. EnGenius Technologies (2017) Wi-Fi beacon frames simplified. Online available: https://www.engeniustech.com/wi-fi-beacon-frames-simplified/
13. Henrik S, Mikko S, Jere K, Pasi R, Bluetooth® 5, Refined for the IoT. Online available: https://www.silabs.com/documents/public/white-papers/bluetooth-5-refined-for-the-IoT.pdf
14. LoRaWAN Alliance, Online available: https://lora-alliance.org/about-lorawan
15. Jacoby D, IoT: how I hacked my home, online available on https://securelist.com/iot-how-i-hacked-my-home/66207/
16. Westerlund M, Leminen S, Rajahonka M (2014) Designing business models for the Internet of Things. Technol Innov Manage Rev 4(7):5–14
17. Rose K, Eldridge S, Chapin L (2015) The Internet of Things: an overview understanding the issues and challenges of a more connected world. Internet Soc (ISOC) 22
18. Tamane S, Solanki VK, Dey N (eds) (2017) Privacy and security policies in big data. IGI Global
19. Vimal S, Khari M, Crespo RG, Kalaivani L, Dey N, Kaliappan M (2020) Energy enhancement using multiobjective ant colony optimisation with Double Q learning algorithm for IoT based cognitive radio networks. Comput Commun
20. Dey N, Fong S, Song W, Cho K (2017) Forecasting energy consumption from smart home sensor network by deep learning. In: International conference on smart trends for information technology and computer communications. Springer, Singapore, pp 255–265
21. Mukherjee A, Dey N (2019) Smart computing with open source platforms. CRC Press
22. Mukherjee A, Panja AK, Dey N (2020) A beginner's guide to data agglomeration and intelligent sensing. Academic Press
23. Mhetre NA, Deshpande AV, Mahalle PN (2016) Trust management model based on fuzzy approach for ubiquitous computing. Int J Ambient Comput Intell (IJACI) 7(2):33–46

24. Babar S, Mahalle P, Stango A, Prasad N, Prasad R (2010) Proposed security model and threat taxonomy for the Internet of Things (IoT). In: International conference on network security and applications. Springer, Berlin, pp 420–429
25. Mahalle PN, Anggorojati B, Prasad NR, Prasad R (2013) Identity authentication and capability based access control (IACAC) for the internet of things. J Cyber Secur Mobil 1(4):309–348

Chapter 2
IoT Use Cases

2.1 Overview

In order to simplify our life, we try to innovate. We make use of technology to make our life easier. IoT and AR will play vital role while in enhancement of IoT services. A few scenarios are studied and presented in this chapter to discover every facet of IoT. For each scenario, study is targeted toward the type of IoT devices needed, method of organizing devices, type communication technology used, services offered, user roles and the mechanism of service advertisement and security aspects. AR with IoT makes people life more comfortable by converging digital infrastructure with physical environment. In this chapter, few promising IoT use cases are discussed. Furthermore, how AR can improve these IoT use cases are also discussed by providing various possibilities of AR to improve these use cases. In this chapter, various sector like e-health, retail, smart home, travel and tourism, smart office are considered. In each sector, one case study is explained in detail and scope of that particular sector describes the possible or in some cases implemented applications of AR.

With the rise in IoT technology, there will be explosion of IoT services through smart devices; IoT is used in many applications to improve human life [1–7]. When IoT is combined with AR technology, it will provide users result with better visualization as well as real-time information. Most important advantage of combining AR with IoT is to cover the gap between real and digital world. Easy access to information and dynamic nature of this technology will result into the products that are beyond human imagination.

2.2   e-Health

Health always has been an area of concern as health issues need urgent attention and delay in getting medical treatment could prove fatal. In health domain, IoT plays a major role to provide facilities that could improve healthcare methods and help to save valuable human life [8–17]. In the literature, tremendous work is done in the e-health domain for functions like ECG monitoring [18], patient tracking [19], smart ICU [20], hospital data management [21] and remote patient monitoring [22]. However, to realize smart hospital, IoT faces many hurdles. The scenario of e-health described below gives the details about how IoT has enhanced medical care methods.

Actors

1.   Doctor (Dr. Shinde)
2.   Patient (Ms. Nisha)
3.   Nurse (Mrs. Mehata).

Services

1.   Hospital map
2.   Route to different departments
3.   Patient registration
4.   Patient admission
5.   Payments
6.   Canteen-related services.

Miss Nisha is a diabetic. She takes an appointment of Dr. Shinde at a city hospital. On her scheduled appointment date and time, a cab is booked from her location to the city hospital (**access control of the location and personal attributes**). When she reaches the hospital, at the front desk, she registers her details and preferences. After registration, she gets the hospital map, showing her destination as Dr. Shinde's Cabin. Hospital infrastructure devices direct her route to the Dr. Shinde's cabin (**discovery of public things and consumption of services**).

As she reaches the cabin, she gets an alert about how much time she needs to wait. After waiting time, she gets another alert on her smartphone for calling her inside the doctor's cabin. After examination, she needs to go to the pathology department of the hospital for a few tests. After the test at the pathology laboratory, Dr. Shinde suggests her for a blood sugar indicator and insulin injector. It is lunch time; hence; she gets a notification of lunch from the hospital cafeteria. In keeping with her preferences (**Preference-based service discovery**), a menu card is flashed on her smartphone. When she reaches home, her health parameters are continuously sent to the hospital database, through implanted/wearable sensors and smartphone. These data are monitored by Mrs. Mehata (**limited access rights assigned**). In case of abnormal readings, Mrs. Mehata gets an alert. Then, a notification of emergency is given to Dr. Shinde. After notification, he changes Nisha's diet plan and insulin level of the injector.

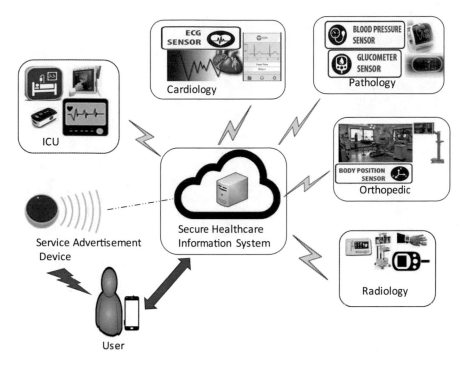

Fig. 2.1 Hospital scenario

The diagrammatical representation (as shown in Fig. 2.1) of the case study is as follows:

Scope for AR

In order to simplify our life, we try to innovate. We make use of technology to make our life easier. Augmented reality will play vital role while in enhancement of healthcare industry.

Sometimes, it becomes impossible for patients to accurately describe their symptoms to doctor. At that time, a wrong diagnosis can lead to a fatal situation. This problem is very common. The main reason behind this is less knowledge about situation or overreaction of the patient.

To tackle this problem, an AR app called EyeDecide was invented. This app makes use of AR to display simulations of the vision. These simulations have been designed for wide array of eye diseases. Via these simulations, patients are accurately able to identify their symptoms and correctly describe it to the doctor. Hence, to help doctors to make accurate diagnosis of the situation.

One of the major parts of the health industry is surgery. In surgery, AR actually helps to save lives. Sometimes, it becomes very difficult to get clear visualization of the tumors or organs. A clear and better visualization always helps surgeons to get the job done. Hence, in this particular situation, a three-dimensional reconstruction of organs or tumors will greatly aid surgeons in visualization that will in turn get the best result for the patients.

Many times, the interns are not able to find the veins of the patients. In this situation, instead of taking extraordinary measures to find the veins, a simple handheld scanner projector can be used. It is projected on the skin and shows the patients' veins. It saves a lot of time and trouble for both doctor and the patient.

Medications are also one of the major parts of healthcare industry. Medications are prescribed to the patients in order to heal and recover faster. AR encourages patient to improve their knowledge about medications that are prescribed to them. AR displays the information about medication in the understandable and memorable form. It has been observed that the a little knowledge about the medication can really help patients to trust that particular prescribed drug. Marker-based AR provides patients with the three-dimensional model of medication, main area which going to be affected with the drug, how that particular medication is going to solve the problem within the body.

2.3 Supermarket

Shopping for grocery items is a tedious and time-consuming task. IoT can make it easier using technologies like RFID tags, readers and sensors. In the literature, IoT is used in the supermarkets for asset management, awareness and recommendation of products, payments [23–25]. The scenario of supermarket illustrated below demonstrates the use of IoT in the retail market.

Actors
User: Mr. Yogesh comes to supermarket for shopping
       Level 1 worker in the supermarket: Mr. Mohit

Services

1. Map of supermarket
2. Route toward different section of supermarket
3. Offers on the items
4. Status of shelf containing items
5. Update of pricing and offers
6. Payment.

Mr. Yogesh lives in a smart home. Hence, information about expiry date and quantity of grocery items is stored and updated at the server. From this information, he gets a list of items to be purchased on his smartphone (**service consumption**).

He is a regular customer of this supermarket. When he enters the supermarket, depending on his previous shopping, few items are suggested to him. As per the list

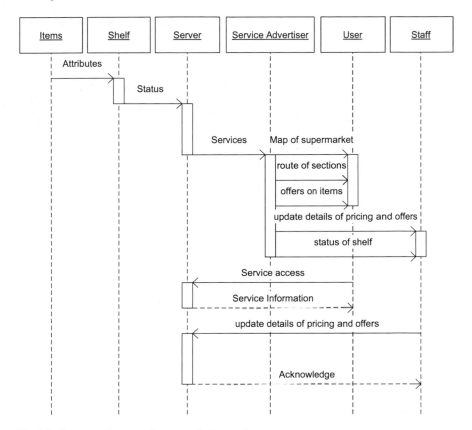

Fig. 2.2 Sequence diagram of supermarket scenario

of items, supermarket trolley suggests him a route to pick up items and supermarket infrastructure broadcasts different offers (**context and preference-based service discovery**).

The shopping trolley calculates the total amount to be paid and transfers the amount details to the cash counter. At the cash counter, the amount is deducted from his account using smartphone and device of the supermarket (**access control**).

Mr. Rohit gets an alert from the supermarket shelf about a fewer number of items remaining on the shelf. From the supermarket database and shelf number, the specific item and quantity required on the shelf is flashed on his smartphone, and then he orders items from the supermarket warehouse. Mr. Rohit gets another alert to change details of prices and offers related to the specific item. He accesses the supermarket database and updates the details about items and offers.

The diagrammatical representation (as shown in Fig. 2.2) of the case study is as follows:

Scope for AR
Grocery shopping is part of the day-to-day life. AR helps and makes this part of life a little bit easier for user. AR permits user to personally interact with the products that are currently available in the store. AR can show the detailed information about the product along with its location. It can greatly help user to find a particular product very easily.

User will also be able to customize color and style of the required product. AR will also be able to provide availability of the same product in the nearby stores along with the information about offers on that particular product.

In the future, all the details that printed on the product will take a new form. It means that the information will adapt to AR in order to gain more profit. User will gain number of benefits and along with that the companies will be able to lessen the gap between printed and digital media. It will also greatly aid companies to understand the mindset of the user. It will also tell companies how users react and respond to their product.

The excellent example of this is Vespa, a famous Italian scooter retailer. This company has linked their own printed magazine using AR app. With AR app, users can scan add and customize their bike according to their personal style along with colors and accessories, etc.

Many international companies have braches all over the world. The main difficulty with this is the language barriers between the continents. In order to sell any product, it is very important to get connected with the user, and if the language is not the native selling of the product becomes that much difficult. AR provides a better solution for this language barrier problem. Google translate AR mode permits users to view any text in 40 foreign languages along with the native one.

If users are able to preview the product with AR before purchase, it will also strengthen the willingness of customer to buy that product. It is also observed that AR saves users valuable time as well as helps them to make decisions.

2.4 Smart Home

In the literature, considerable work is done in the field of smart home for controlling home devices remotely [26, 27], efficient energy consumption [28] and home security [29]. Generally, smart home applications are developed by IT people based on requirements of the end user (IT or non-IT person). Hence, it is difficult for the user to add/update access rights for devices. To overcome this, Hafidh et al. [30] proposed the smart home methods where non-IT user too can set access rules; i.e., end user development facility is provided. In the smart home, devices/sensors are connected to the Internet and user can access it ubiquitously. Further, sensors/devices learn from information sensed by them and take actions; i.e., if allergic particles are sensed in the home garden or outside the home, then windows of the home will automatically close.

Actors

1. Home owner: Mr. Joshi
2. Kid at home: Atharva
3. Guest: Mr. Bhave.

Services

1. Main door access
2. Adjust lights brightness, switch on and off
3. AC/heater on/off
4. Control of TV and other electric appliances

Mr. Joshi lives in a smart home. Every morning, he wakes up at 6 a.m. and exercises at his home gym. Depending on his body weight and health parameters, the gym equipment starts functioning. According to the brightness outdoors, the home lights regulate. As he stops exercising, the bath tub starts filling with hot water, and it is ready for his bath (**context aware service consumption**).

His smart refrigerator suggests the recipe for breakfast depending on vegetables and ingredients available in the fridge. When he leaves home for office, the major electronic equipments automatically switch off, e.g., washing machine, gym equipment, dish washer as only Atharva, his kid, stays at home. Atharva can use other home equipment according to limited access use (**limited access right**).

Furthermore, Mr. Bhave comes to stay at Mr. Joshi's place for a few days. Access of the main door, gym, car, some few electrical equipment in the home, lights, AC/heater of guest room are given to Mr. Bhave during his stay (**access delegation**).

Scope for AR

Smart home combined with AR tech can help resident of that house in numerous ways. If resident of the house is thinking of the remodeling, then before starting with remodeling, it will be beneficial to see the plumbing and electricity layout of the house. Otherwise, it can result into a disastrous situation. These layouts are available, but most of the times, they are very difficult to read. Hence, AR can help you with the visualization of all the plumbing and electricity layout of the house. Now, a user can safely remodel the house.

Same concept can be applied in case of the interior decoration. If a user want to decorate say living room, and all the supply for the decoration is ready but user is confused on which thing will go where. In this case, AR can help user with actual visualization of the living room and how each thing will look if placed at a certain position. User can also try out different scenarios before finalization of a certain look.

Wardrobe is important part of the house. Everyday, there is a huge struggle to choose outfit for say to office or to an important function. With AR, user can be presented with all the options that are available currently in user's wardrobe. From those options, user can select an outfit and be on the way. This will save a lot time.

Elderly people need support in multiple activities around the house. AR combined with IoT can help elderly people with the navigation around the house. With AR, they can safely navigate.

Many times user can forget a particular food item after putting it inside the fridge. With AR, you will be able to see which food items are currently available in the fridge. Also, AR combined with IoT will be able to tell user whether that particular food item is in edible condition or not.

2.5   Travel and Tourism: Airport

Use of IoT technology in the airport scenario is proposed in the literature. IoT could be used to provide different services related to flight and luggage status [31, 32].

Mr. Rai is a senior salesman of a world-renowned company. His job necessitates frequent travel. The scenario given below is one of the business trips of Mr. Rai.

Actor Mr. Rai

Services

1. Map of the airport
2. Waiting time at security area
3. Flight status
4. Gate Numbers
5. Route from gates
6. Menu cards of cafeteria
7. Luggage alert
8. Map of city
9. Nearest bus stop and bus schedule.

Mr. Rai enters the airport and gets an alert on his smart device showing the different services available at the airport (**discovery of public things and services**), e.g., a guided map of the airport, the expected waiting time in the security check area, airline services, etc.

He chooses to check the expected waiting time and is informed by a thing in the airport that on an average it takes half an hour to clear security (**service consumption from public things**).

At the check-in desk, another alert informs him that due to a technical snag his flight is delayed by a couple of hours and lunch e-vouchers are provided by the airline. (**Alerts based on personal information made available to the airline thing**).

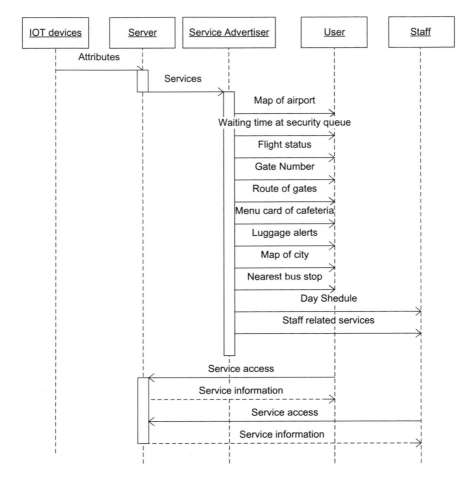

Fig. 2.3 Sequence diagram of airport scenario

At the destination airport, he gets an alert about the location from where he needs to collect his luggage. His luggage is equipped with RFID tags and sensors. As luggage comes on a conveyor belt, he gets an alert which negates the need to check the belt continuously. Instead, he can rest and relax after the long journey. He collects his luggage and moves toward the exit gate.

At the exit gate, he gets an alert of the city map and bus stop map on his smartphone. Since he is new and unfamiliar to the city, with the help of his current location, the bus stop services provide him with information about the nearest bus stop and bus ticket service. Based on this information, he buys a ticket for his destination (**access control to approve payment**).

The diagrammatical representation (as shown in Fig. 2.3) of the case study is as follows:

Scope for AR

Many countries have tourism as largest contributor in their domestic economies. Technology at each step has always revolutionized tourism industry. This time, it is especially critical when the bigger parts of travelers are millennials. AR has wide-range applications in the tourism industry. AR can help travelers to enrich their local experience by providing all the helpful information via AR app. This app will be able to provide assistance in case of navigation. Along with that when traveling many times, language is the biggest barrier. AR app can provide an efficient language translator. Also, if you need some local assist or guide, it can also be arranged by the app.

When travelers land in new city, the feeling of being in a new city is exciting. But along with that feeling, it is also nerve wrecking and overwhelming at the same time. It is because you are in totally unfamiliar place with no clue where to go next. At this time, if you have travel app combined with AR you can point at the transportation object, which will provide directions, routes, last stop along with next possible place of interest.

With the help of AR, you can transform metro maps or other items into more interactive interface which will have wide language support. For example, anAR app called tunnel vision does this with the New York subway. This app provides you with scheduled data along with some other information.

Food is also very big part of tourism sector. Consider you are in a brand-new restaurant and you are not aware of the dishes on the menu. At this time, if you have an AR app which will provide you a 360° view of the particular dish it will be more beneficial. Along with that you will also be able to know the portion size and ingredients list in detail for a particular dish.

Exploring a new city, its tourists attractions is always exciting. But along with that if you have an AR support to see how that city's landmarks were developed over the time using three-dimensional modeling then that experience will be extraordinary.

In the last decade, museums which are culturally, historically rich have seen a decrease in visitor. It is because museums have not taken any support of latest technology to alter the legacy system. AR can be of a great assist to the museums all over the world. Every piece of art from the museums can be turned into augmented models. The excellent example of this is the National Museum of Natural History. The museum have launched skin and bones app which helps people to witness the live representation of animals which have been extinct over the years. For this live representation, it makes use of the skeletons of the animals which are there in the museum already. Theses museums can really reinvent the concept of the museum with the help of the AR. AR can help in creation of visual tours which are more interesting than older walk and talk method. These tours will be able to provide all the information to the travelers in a way that it will be in their memory for a longer time. This can a really fascinating experience for the travelers.

AR can also be used in following things:

- Hotel tours, booking
- Easily accessible travel information
- Advanced navigation.

2.6   Smart Office

Nowadays, industry offices have become smart. Devices in offices and in industrial plants can be accessed anytime from anywhere [33–35]. With smart devices and improved machine-to-machine communication, systems can run with less user intervention, thereby saving and controlling costs. The scenario of such a smart office is illustrated below.

Actor
Project Manager: Mr. Sen
  Project Member: Mr. Prabhu from a different branch of the firm

Services

1. Day Schedule
2. Conference room availability
3. Project status
4. Cab booking
5. Environment control
6. Breakfast/lunch menu.

Mr. Sen works in a smart office which has a few branches distributed in the city as well as globally. He is the project manager and needs to communicate with project members of different branches. He has to go through very busy and tough days at the office. The scenario described below is of one such day.

When he enters the office gate, he gets an alert for his day's schedule. A breakfast menu is recommended consistent with his preferences. Depending on the weather, the temperature of the cabin is maintained, i.e, AC/heater on/off or temperature adjustment (**context aware service consumption**). As per project's progress, he can convene an emergency meeting if necessary. He uses conference availability service to book conference room for a meeting (**services available from infrastructure**). At the time of the meeting, the conference room is set to meeting mode; i.e., appropriate light brightness, projector screen sliding down and Google spreadsheet are shared among all members who are participating remotely.

During the day's schedule, if he needs to go to another branch for a visit to the project plant, then accordingly the smart phone books a cab (**access control**). When Mr. Sen is at the exit gate, the cab is already waiting for him.

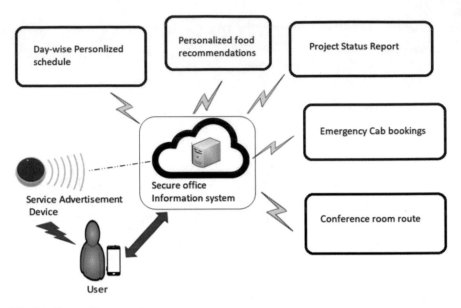

Fig. 2.4 Smart office scenario

Mr. Prabhu, who comes to Mr Sen's office for a few days to work on the project and gets access to devices and office services related to his work. (**limited and controlled access rights**).

The diagrammatical representation (as shown in Fig. 2.4) of the case study is as follows:

Scope for AR
Industry offices many times are distributed among multiple floors or buildings. In this situation, it can become very difficult for user to learn and memorize location of every room in the building. AR can easily help user try and locate a certain room along with the directions in more interactive manner.

AR can also be used in the presentations so that user can give confidence to stake holders that he/she have the complete control and ability to finish that project. AR can greatly impact the critical presentations.

Also, before manufacturing process is initiated, it is necessary to be absolutely sure that product does not have any type of the bug in it. AR can assist in the complete visualization the product before manufacturing. AR will be able to show you the potential bugs that can harm the product.

2.7 Conclusions

AR combined with IoT bridges the gap between real world and digital world by actual interaction with the real world in real time. Augmented reality devices (mounted/portable) collects real-time data from the current environment. According to what user is currently doing, only information needed for current work is intuitively displayed. AR combined with IoT can be used for wide array of applications. It can be used for handling of maintenance issues in remote localities where visibility is compromised or high temperature situations. It can also be used for hazardous conditions. It can also help business ventures, health care and other domains you can think of. Sky will be the limit with this kind of applications.

References

1. Sarowar MG, Kamal MS, Dey N (2019) Internet of Things and its impacts in computing intelligence: a comprehensive review—IoT application for big data. In: Big Data analytics for smart and connected cities. IGI Global, pp 103–136
2. Pal AK, Banerjee S, Dey N, Sengupta D (2018) IoT based home automation. In: 2018 3rd international conference for convergence in technology (I2CT), Apr 2018. IEEE, pp 1–6
3. Dey N, Bhatt C, Ashour AS (2018) Big data for remote sensing: visualization, analysis and interpretation. Springer, Cham. Dey N, Tamane S (eds) (2018) Big Data analytics for smart and connected cities. IGI Global
4. Sarkar M, Banerjee S, Badr Y, Sangaiah AK (2017) Configuring a trusted cloud service model for smart city exploration using hybrid intelligence. Int J Ambient Comput Intell (IJACI) 8(3):1–21
5. Shinde GR, Olesen H (2018) Beacon-based cluster framework for internet of people, things, and services (IoPTS). Int J Ambient Comput Intell (IJACI) 9(4):15–33
6. Schaller A, Mueller K (2009) Motorola's experiences in designing the Internet of Things. Int J Ambient Comput Intell (IJACI) 1(1):75–85
7. Dey N, Mukherjee A (2018) Embedded systems and robotics with open source tools. CRC Press
8. Sengupta D (2020) Taxonomy on ambient computing: a research methodology perspective. Int J Ambient Comput Intell (IJACI) 11(1):1–33
9. Bhatt C, Dey N, Ashour AS (eds) (2017) Internet of Things and big data technologies for next generation healthcare
10. Dey N, Hassanien AE, Bhatt C, Ashour A, Satapathy SC (eds) (2018) Internet of Things and big data analytics toward next-generation intelligence. Springer, Berlin, pp 3–549
11. Dey N, Ashour AS, Bhatt C (2017) Internet of Things driven connected healthcare. In: Internet of Things and big data technologies for next generation healthcare. Springer, Cham, pp 3–12
12. Elhayatmy G, Dey N, Ashour AS (2018) Internet of Things based wireless body area network in healthcare. In: Internet of Things and big data analytics toward next-generation intelligence. Springer, Cham, pp 3–20
13. Dey N, Ashour AS, Shi F, Fong SJ, Tavares JMR (2018) Medical cyber-physical systems: a survey. J Med Syst 42(4):74
14. Hassanien AE, Dey N, Borra S (eds) (2018) Medical big data and Internet of medical things: advances, challenges and applications. CRC Press
15. Dey N, Mahalle PN, Shafi PM, Kimabahune VV, Hassanien AE, Internet of Things, smart computing and technology: a roadmap Ahead

16. Dey N, Ashour AS, Shi F, Fong SJ, Sherratt RS (2017) Developing residential wireless sensor networks for ECG healthcare monitoring. IEEE Trans Consum Electron 63(4):442–449
17. Chluski A (2018) Impact of building human capital with support of information technology on efficiency of hospital activities. Int J Ambient Comput Intell (IJACI) 9(2):1–15
18. Nurdin MRF, Hadiyoso S, Rizal A (2016) A low-cost Internet of Things (IoT) system for multi-patient ECG's monitoring. In: 2016 international conference on control, electronics, renewable energy and communications (ICCEREC), Bandung, Indonesia, pp 7–11
19. Laplante NL, Laplante PA, Voas JM (2016) Stakeholder identification and use case representation for Internet-of-Things applications in healthcare. IEEE Syst J 1–10
20. Ahouandjinou A, Assogba K, Motamed C (2016) Smart and pervasive ICU based-IoT for improving intensive health care. In: 2016 international conference on bio-engineering for smart technologies (BioSMART), Dubai, UAE, pp 1–4
21. Thangaraj M, Ponmalar PP, Anuradha S (2015) Internet of Things (IOT) enabled smart autonomous hospital management system—a real world health care use case with the technology drivers. In: 2015 IEEE international conference on computational intelligence and computing research (ICCIC), Tamil Nadu, India, pp 1–8
22. Archip A, Botezatu N, Serban E, Herghelegiu P-C, Zala A (2016) An IoT based system for remote patient monitoring. In: 2016 17th international Carpathian control conference (ICCC), Slovakia, pp 1–6
23. Gonzalez-Miranda S, Alcarria R, Robles T, Morales A, Gonzalez I, Montcada E (2013) Future supermarket: overcoming food awareness challenges. In: Proceedings—7th international conference on innovative mobile and Internet services in ubiquitous computing, IMIS, Taichung, Taiwan, pp 483–488
24. Jalkote V, Patel A, Gawande V, Bharadia M, Shinde GR, Deshmukh AA (2013) Futuristic trolley for intelligent billing with amalgamation of RFID and ZIGBEE. Int J Comput Appl 5:0975–8887
25. Guo H, Li J (2014) Research on the application of intelligent campus supermarket system-based on the Internet of Things (IOT) technology. In: 2014 seventh international symposium on computational intelligence and design, pp 390–394
26. Shinde G, Olesen H (2015) Interaction between users and IoT clusters: moving towards an Internet of People, Things and Services (IoPTS). In: World wireless research forum meeting, vol 34
27. Gowrishankar S, Madhu N, Basavaraju TG (2015) Role of BLE in proximity based automation of IoT: a practical approach. In: 2015 IEEE recent advances in intelligent computational systems (RAICS), Kerala, India, pp 400–405
28. Salman L et al (2016) Energy efficient IoT-based smart home. In: 2016 IEEE 3rd world forum on Internet of Things (WF-IoT), Reston, VA, USA, pp 526–529
29. Fernandes E, Rahmati A, Jung J, Prakash A (2017) Security implications of permission models in smart-home application frameworks. IEEE Secur Priv 15(2):24–30
30. Hafidh B, Al Osman H, Arteaga-Falconi JS, Dong H, El Saddik A (2017) SITE: the simple Internet of Things enabler for smart homes. IEEE Access 5:2034–2049
31. Ghazal M, Ali S, Haneefa F, Sweleh A (2016) Towards smart wearable real-time airport luggage tracking. In: 2016 international conference on industrial informatics and computer systems (CIICS), Sharjah-Dubai, UAE, pp 1–6
32. Ye-Won L, Yong-Lak C (2015) Proposal for air-baggage tracing system based on IoT. In: 2015 9th international conference on future generation communication and networking (FGCN), Jeju Island, Korea, pp 25–28
33. Olivieri AC, Rizzo G, Morard F (2015) A publish-subscribe approach to IoT integration: the smart office use case. In: 2015 IEEE 29th international conference on advanced information networking and applications workshops, Gwangju, Korea, pp 644–651

34. Cho K, Kim SH, Kang B, Jang SM, Park S (2017) Intelligent office energy management system by analysis in Hyper-connected-IoT environments. In: 2017 IEEE international conference on consumer electronics (ICCE), Las Vegas, Nevada, pp 289–290
35. Nelis J, Vandaele H, Strobbe M, Koning A, De Turck F, Develder C (2015) Supporting development and management of smart office applications: a DYAMAND case study. In: 2015 IFIP/IEEE international symposium on integrated network management (IM), Ottawa, Canada, pp 1053–1058

Chapter 3
Context Management in IoT

3.1 Context and Context Management

Internet users' are carrying several smart devices with them and moves with it in this rapidly changing world of Internet of Things. These smart devices and IT environment have brought up revolutionary changes in the user's life, businesses, etc. Existing devices and inclusion of computational devices are making a big difference in communication with other entities (user, devices, etc.). This revolutionary combination is executing productive tasks that reduce human intervention with computers. This has led to a vision in 1991 by Mark Weiser [1] who marked as the vision of pervasive computing [1]:

> The most profound technologies are those that disappear. They weave themselves into the fabric of everyday life until they are indistinguishable from it.

The abstract above vision is to create an environment that has the capability to communicate and compute with well-balanced human integration. Looking at the progress in hardware and technology happening now, the infrastructure required for pervasive system is available and we could pursue the vision of pervasive system [1].

Pervasive system provides services to the Internet users that interact with pervasive environment. The interaction consists of Internet users, smart devices, applications and environment. The devices are not limited to desktops or laptops. The users are moving from desktop version devices to mobile version devices who are capable of interfacing in diverse environments. The users can access any service from any device in any environment. Using smart devices in pervasive system, the basic and advanced services are offered to users according to the users' behavior in a dynamic environment. These physical devices (sensors, human–computer interfaces, etc.) are responsible for providing basic services in pervasive system. To provide advanced services, these devices become services that gain and interpret valuable information like user's current task, physical location, social position, as well as concerned information in a pervasive system.

G. R. Shinde et al., *Internet of Things Integrated Augmented Reality*,
SpringerBriefs in Computational Intelligence,
https://doi.org/10.1007/978-981-15-6374-4_3

Having said before, the upsurge use of smart, tiny and wearable devices has a quick access to a mobile network that generates a large amount of user and location data. This valuable data should be made available to users at the right with the help of an interface. The new interface is augmented reality (AR) that permits digital data to interweave with physical places [2]. Having said in earlier chapters, AR allows overlapping of digital space into physical spaces using graphical augmentation which are interactive in nature during real time. This makes the empowerment of AR users to have instantaneous interaction with physical and digital world. AR applications are not restricted to the laboratory environment, and it has been observed use of AR in many real-time places or domains [3].

The applications of AR used for short times and have a single purpose. However, recent development in mobile networks and availability of useful data can provide a continuous experience of AR. Hence, the new interface of AR is dealing with information in digital and physical world continuously. This continuous pervasive system should adopt current and changing information that characterizes any entity in this computing world.

In AR experience, the behavior of user is vital, and it changes along with user's surrounding environment, nearby users and devices. Hence, the universal interface of AR should be capable to understand user's behavior in different contexts. It should be aware about the environment within which users and devices are progressing or changing [4]. AR should identify those changes in a dynamic environment, in particular, location (temporal and spatial), physical devices, etc., and accordingly, the user should receive desired services. Hence, the two vital components are users and the environment, and their interaction is represented in Fig. 3.1 [5].

The interaction model represented in Fig. 3.1 shows two important components as users (devices and application) and environments (electronic and physical). Using devices, the users are accessing applications in a surrounding environment. This interaction between users and environment generates valuable information that stands

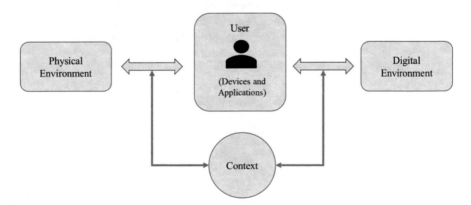

Fig. 3.1 Interaction model between user and environment

basis to define the context. The valuable information can be user's location, preferences, time, etc., during browsing of news channel or buying item online. Let us consider a few applications to understand pervasive system, context and its management.

In a pervasive system, the context management system provides many applications that involve context. To understand context management system, here three context-based application scenarios are discussed.

- **Room Temperature Control Application**:
  In this application, the temperature in a room is checked, and accordingly, the actions are executed. Depending on the temperature in a room, the application would turn on/off room heater or air conditioner. Also, the application can decide to open the room window if temperature is high, or to achieve energy, the application can pull up the window curtains.
- **Phone Call Management Application**:
  A mobile phone has several modes (ring, silent, vibrate, etc.). In order to maintain a user's state (meeting, traveling, sleeping, etc.), the phone call management should act accordingly. Based on the importance of a call, the application should determine whether the phone should ring, silent, vibrate or route to another phone, etc. If the user is sleeping or he is in an important conversation with others, the application should determine the users' state and should execute desired action. To make a decision, the required information could be user's devices in use (mobile, computer, TV, etc.), users' activity (sleeping, meeting, etc.), other entities (users, devices, etc.), state of environment (light, temperature, humidity, etc.).
- **Car Music Control Application**:
  The application could play the music in the car based on user's states/situations. If a user is driving "alone," then the application would play his "favorite" songs. If the user is driving with "friends," then the "party" songs would be played by the car music application. In other case, if the user is driving, then "family/devotional" songs would be played. If the user is with kids, then "kids' songs" would be played. In this application, the user's preferences and situations are identified by the applications and similarly the songs are played.

Learning from Application Scenarios
Above application scenarios reveal many key perceptions of pervasive system. First application reveals "cross-platform" of pervasive system. In room temperature, controller application understands the condition on a room, and accordingly, the devices are instructed to perform the desired action. The action could be turned on the heater or air conditioner. The room can office room or can be a room in a house. This application also reveals the prominence of understanding of context. In this case, context could be location, temperature, time, etc.)

A second application shows the significance of "awareness" of user activity or state. Based on user's state (sleeping, meeting, etc.), the application should handle the calls made to the user. The application dynamically checks the user's status and decide to determine if it must ring, silent, vibrate or redirected to another number. The application reveals the understanding of user state and behaves accordingly.

Third application shows the importance of "pro-activeness" functionality. Based on the user's mood or situation, the music is being played. The car music application understands the situation of the user (alone, with friends or family) proactively, and the desired action (song list) is played. This application also reveals additional information to understand the context as (location, preferences, users, etc.).

Hence, the prime intention of pervasive system is to enable "smart" devices that are capable of creating of "smart" environment. This "smart" environment should be able to collect, send, store and process surrounding information which understands the context and related activity. In pervasive system, situational and circumstantial information is called as "context." Hence, the pervasive system should identify the context, and context-based services should be given to the user. Let us understand context, context metadata and context management in detail.

- **Context and Its Types**

Communication between users and environments reveals information in a pervasive system. Having said before in Fig. 1.1, the interaction between users (devices and applications) and environment (physical and electronic) describes the information related to the communication between these two important components, and this is called as context. A context can have an impact on user's behavior with AR systems. The factors like human (personal or social), environment (physical or digital) and system (input, output or state) can be vital to understand as context sources to AR systems. User's attributes, preferences, choices and interaction among other user's results into valuable information to understand user behavior in AR system. The interaction between user and environment is important to visualize "scene" in AR. This scene describes the information flow between users and the environment (physical or digital). The infrastructural components like system configuration, input and output devices (touch screen, number of displays, mouse inputs, etc.) which are connected to AR system will reveal context information. The meaning of context from a dictionary is to bind together or the part or parts that are associated with each other [6]. This could be related to the object or situation or environment. Therefore, this generic form of definition needs to be specific and detailed.

Schilit et al. [7] have given a definition of context in 1991 as the identities of objects and people along with their location. To add further, context also defines the changes of objects. To offer more customized services to users, user's and device's location information are useful context. The dynamic change in user location should be adopted. In 1997, Brown et al. [8] extended the definition of context by including time, day, identity, temperature, season, etc. Based on the user's preferences, the recommended system was developed. Hence, combining identity and location with additional information (time, day, temperature, etc.) will produce additional information to understand user's requirement and provide desired services to users.

The detailed definition of context has been given by Dey et al. [9] in 1998. It is a collection of information describing the situation of an entity. The definition has emphasized on identity, location, date, time and orientation. The conception of context is not world-wide, but it is concerned with some situation. For example, it could be an users' current task, user location, etc. Partial information that characterizes the situation is also a context. To continue with this definition, user's emotions and attention are also considered.

Looking at the definitions so far, it is unclear that whether a context should include user interest and their choices? Few definitions say that context is an environment or a situation. However, it does not reveal whether it is user-centric environment or application-centric environment. Since, there are definition that says context is a user's environment [8, 10] or the application's environment or both [11], [12]. On the other hand, current situation of the user is termed as context [13].

The precision of the context definition can be seen by Dey et al. [9] and Schilit et al. [7] that has focused on objects and users along with their valuable attributes like location, nearby resources, surrounding environment, etc. In their definition, the execution environment was focused which is likely to change. The executing environment can be a user or physical or computational as discussed here [9, 14]:

- **User Environment**:

In this definition, there are many parameters that could be context that includes user location, time, date, neighboring objects and people and social state.

- **Physical Environment**:

In this definition, the external entities/situations are considered. This includes lighting situation and noise level in a physical environment.

- **Executing Environment**:

In this definition, all the components involved in computations like I/O devices, networking devices and its connectivity, processing units, execution cost, etc.

Hence, the definition is covering several states of users that include physical, informational, emotional and social state. To summarize all the discussion on definition of context, Dey et al. have given thorough definition of context as [9]:

> any information that can be used to characterize the situation of entities (i.e. whether a person, place or object) that are considered relevant to the interaction between a user and an application, including the user and the application themselves. Context is typically the location, identity and state of people, groups and computational and physical objects.

This definition is more accurate and reveals so many information that correlate with all the definitions mentioned in this section. Using definition, an understanding of user and user's activity state is simple. This will help to identify user requirements in a more accurate way. Based on the context and source of context information, it is easy to know the ways to observe the information. Having a discussion on context definitions, a user-centered context can be represented in Fig. 3.2.

A context will also assist to methods of gathering information and helps in management of large-scale information of objects in a given environment. Once the large-scale information is gathered, the challenge is to decide the usefulness of information.

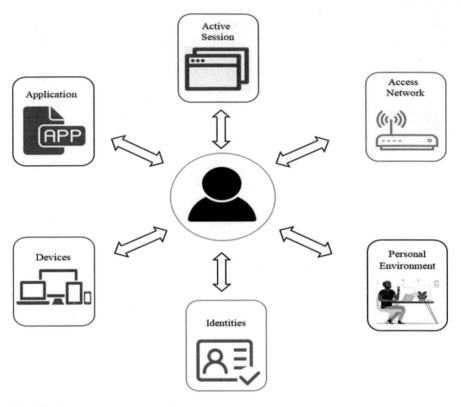

Fig. 3.2 Context—a user perspective

In this issue, the context will help to understand the importance of gathering information. The adaptive feature of context will help to customize and apply application/services to the user as per the surroundings/environment. For example, car music control application that understands the user state or environment before playing any songs/music.

To identify and interpret the context, the model should be able to understand both input and output. In input, the model should consider implicit as well as explicit inputs. Though the definition of context is revolving around situation or environment, there is a requirement to know context acquisition methods. The method to acquire information could be manual or automatic or combined. The automatic way of acquiring information is recommended in the ideal case; however, in the real case, user intervention is required whenever an automatic information gathering is not possible. The definition also reveals the context and its types.

It is vital to understand the information flow during the communication between two entities (user to user, user to machine and machine to machine). Since, this information will help to understand the context and its category. Considering any application in a pervasive system, the context could be location, user interest, connection link in wireless communication, knowing the user's schedule, user requirement and

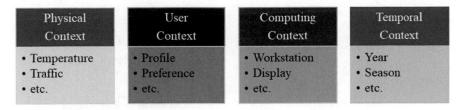

Fig. 3.3 Categorization of context

preference, etc. Looking at these context information, it is obvious to see diversity in context information. This diversity helps to define context types. At the outset, the vital entities in pervasive system could be users, locations and objects. An individual user or a group of users could be one entity. Based on user's mobility, the location could be static or dynamic in an environment. The geographical places could be a room, home, apartment, streets, offices, etc. The objects could be hardware- or software-based components.

In general, context is categorized as primary and secondary context [15]. Primary context is information collected from devices directly. Example of primary context is information generated by sensors. This information is without any blend with other entities in an environment. Secondary context is inherited from primary context. For example, mobile number or email ID is secondary context, considering the person name as primary context. The context is classified into four types [5] and it is represented in Fig. 3.3.

- **Physical Context**:
  In this category, physical values or entities are involved. The context deals with the values coming from a user or device or both. The examples of physical context include temperature in a room, traffic situations in an area, noise levels, etc.
- **User Context**:
  In this category, user information is considered. User and user information are included in this category. The example includes user profile, locations, preferences, etc.
- **Computing Context**:
  In this category, the information related to communication and computing is included. The examples of computing context are network connectivity, communication bandwidth, computing sources like workstations, printers, etc.
- **Temporal Context**:
  In this category, the context is a temporary information with respect to time. This include year, month, week, day, date and time. Season and its information are also considered as temporal context.

The categories of context are useful to execute desired actions in a pervasive system. Identification of nearby users or objects is nothing but proximity selection. Considering the dynamic nature of pervasive system, the alteration of existing component is another action that requires context knowledge and its type. Change

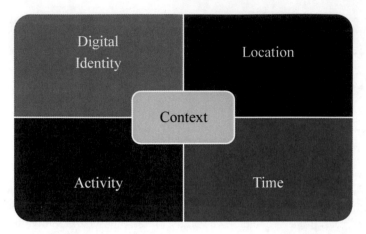

Fig. 3.4 Context classification

in the context or change of the rule is also another action or operation that will be carried out if the context is known.

Another classification of context and context information is given by Dey et al. [9]. The categorization involves digital identity, location, activity, time, and it is represented in Fig. 3.4.

A real-world entity is electronically represented as digital entity. It represents "who" is involved in the digital environment. For a particular entity, there are several attributes (measurable and distinct property of an entity). For instance, user's information (personal, location, behavioral, etc.), user preferences, personal resources, social connects, etc., could be attributes and collectively called as digital entity. An identifier is an attribute or a collection of attributes that identifies a digital entity. In offline mode, the identifier could be a passport or driving licenses, etc. On the other side, the digital identifier could be signing details (user name and password), digital certificate, etc.

Using orientation and elevation, the positioning of an object is determined which is called as a location of an entity. Using location information, the relation between entities are identified. The relationship could be a co-relation, proximity or containment. Location defines geographical area (latitude and longitude) along with the radius. There are two location states of an entity, i.e., active or inactive. There are many location providers like GPS, Wi-Fi, cell towers, etc. Another important context is time that helps to determine a moment or a situation. The historical information of an entity is leveraged using time. The time comes with other context information like timestamp, time span, etc. The combination helps to identify more information which has more relevance to each other. For example, the users' behavior over a given time are identified and respective patterns are generated.

The essential characteristics of an entity are also vital that could help to identify the activity or the state of an entity. A room temperature, light level in a party hall, noise level in an event could be treated as an activity. This activity could be associated with a user. For example, a user is writing (an article), listening (music), chatting (with friends), a group of users attending a meeting or devotees in a temple, etc. Another example could be a state of the application, downtime of CPU, status of file uploaded, etc. In AR, attributes of digital data and physical environment can be combined to identify the relevant context required for interfacing and how to address context awareness.

- **Context Management**

To identify the context and leverage context-based services in AR, there is a requirement of context-aware systems. Some of the context-aware systems have been proposed [16, 17]. These methods are built based on context representations and other value factors. The factors could be acquisition, interpretation, processing, representation and delivery of context. The context-aware management system also considers the meaning of context in heterogeneous platforms. This will help to implement various components that work in the same context. In general, the administration of context is termed as context management. Context management deals with storing of context and helps to provide access, updates and a comparison of the knowledge. Context management encompasses structuring of context, its accumulation and administration. It considers the context and their relevant parameters, the relation between these parameters and sources of information. Context management also includes adaptive behavior in detail. Context management is essential in AR that supports development of context-aware applications covering basic required functionalities. The basic functionalities are context (i.e., collection, management and its' dissemination). These functionalities are ensured if context information, its quality and relationship among context data are considered and represented using context model.

To support and implement the context management system, context modeling (context model) plays a vital role [18]. A model is a way to represent something imaginary or real. A context model is required to represent and store context data in an automatic and computerized way. Context model is used to represent, structure and maintain a context. In a context-aware system, description of context information is given by context model. The interface and behavioral description of context is considered in context modeling [19]. Context management system will be efficient if the context metadata is briefly elaborated and represented. In a pervasive system, the context management is driven by the context information. Therefore, it is important to get correct context information from context sources (like sensors) to make vital decisions by a pervasive system. For example, if there is an error value generated by a sensor, the context will have error. This error-based context will generate undesired services from a pervasive system and will be delivered to users.

As pervasive systems are heterogeneous in nature and behaves dynamically; the context information and its metadata are periodically required [20]. The change in the context information is rapid, and the time intervals of updates are really short. The context type defines the time interval for updates. The context information should be updated and correct. To have detailed understanding of context, its freshness and certainty, additional information about context information is required. This additional information is called metadata of context information. The representation of context metadata is based on context model. The metadata could be source of context, type of context, attributes of context, etc.

The categorization of context management system (CMS) is as below [21]:

- **Domain-Based CMS**:
  In this CMS, the context is collected and handled according to the domain or specific application. The domain could be smart parking, smart home, smartphone, etc. The context related to a domain is collected, inferred and provided.
- **User-Based CMS**:
  Here, the context is gathered, stored and handled as per the user in single or multiple domain. The context could be users' information when he is in smart home or smart office.
- **Interaction-Based CMS**:
  In this CMS, the user interactions are considered. The users could be in the same domain or different domains. All context generated through user interaction is gathered and handled in this CMS.

The context management system consists of several phases, including accumulators, discovery, provider, observer, ontology reasoner and context query. The context management system deals with different devices and read context information. In context accumulator, the contexts are identified based on the application or scenario. This request is initialized and communicated to context discovery. The context discovery sees the type of application and asks the context provider to deliver required context with the help of context reasoner.

Context management system (CMS) consists of mainly two parts [22]. The first part is context modeling tool and the second part is content management system. Figure 3.5 represents the context management system. Each component is discussed further.

The behavior of a system and well-defined sensors are integrated in the source code via Context Toolkit [23]. In 1995, Schilit et al. [24] presented a system architecture that gives an easy way to build context-aware mobile computations. In 1999, the Context Toolkit developed to make a reusable solution that can handle contexts [25].

Context Toolkit an extension to the JAVA programming language that helps to implement context-aware applications. The ready-made software components are available in those libraries which hide the technical details from the user as well as software developers. This toolkit acts a facilitator to implement and install interaction context-aware services [26]. This toolkit encompasses many services related to context like collection, storage, distribution of context information, etc. The procurement of context and its delivery is handled through a widget called context gadgets. However, the toolkit does not have context model and has a lack of support of dynamic changes in an application.

In sensor layer, the software objects receive data, and cleansing of data is done followed by integration of sensor values that are passed to semantic layer. In semantic layer, each context attribute is received from the sensors. Each context attribute is

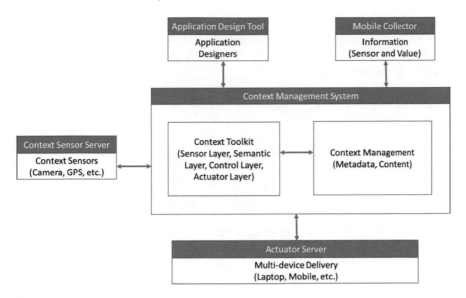

Fig. 3.5 Elements of context management system

connected to one or more sensors. It means that sensors act as information sources, and context attributes acts as information interpreters. Further, each sensor value is mapped to a meaningful value (semantic mapping). For example, time value received from sensor represents "time of day." Also, context attributes are derived from more than one sensor. For example, time and position together will give speed. Each context attribute is stored in one of the three ways. This includes fixed time interval, after a change in the event or explicit input. The change event is based on the context values and the information received from the sensor layer. The context change events are sent to control layer. This event is generated as soon as there is a change in the context values. Based on the events, the triggers are initiated that adhere to the rules of the system. The actuator or indicator layer calls actuation server to initiate the desired application. This could be executed on a single device or on multi-device.

For a design of an application in AR, there is requirement of design tool that creates various panels. The panels are based on sensors being used; attributes are received; it is modeling and controlling, followed by user queries and actuators. The sensor panel gives a leverage to the administration to deal with sensors. The addition and removal of sensors are managed using this panel. Each sensor is assigned a unique ID. On the similar line, the attribute panel helps the administrator to handle attribute (addition and removal of attributes). Each attribute is identified using its unique name. The change of application (actions to change the behavior of the application) is done by actuator panel. Once the basic elements of context-aware applications are well-defined, the modeling step can begin. In this step, context modeling is used. Here, the domain, context and its types are defined. Also, allocation of context attributes and sensors are made. For example, time sensor will deal with time (time of day). This step also includes the metadata definition about the context. In queries panel, preconditions and qualifiers are prepared that will control the behavior of the application. Based on the context attributes, the situation is analyzed. The operations and operands are represented using a tree data structure. The systems' behavior is specified by the control panel.

The context data and its explanation are a distinct task. This could be done using contexts of use. A mobile collector is a support to users to see context information together with content data. Once a content provider gives content with reference to the context, the mobile collector will act as a tool to show contextualized content. The interface of this tool will help the users to add, remove, browse and search contextualized content. Current context device and the sensed values are depicted using this tool. Whenever there is a change in context value, the updates are shown in this tool. Using this tool, a user can freeze current context (values) and can link between contexts.

An adapted browser called content player runs on portable devices like PDA. If a sensor is connected to mobile device, then the sensor's values received from the sensors are sent to the server using content player. On the server side, the sensor values are read and interpreted to determine the desired behavior of the mobile device. The behavior is based on contents from content management systems with reference to the current context. Further, the server sends the content accordingly. The browser is refreshed to see the latest content. In this regard, context toolkit is required to understand the behavior and content in AR system.

3.2   Context-Aware Middleware

In general, a middleware is a software program which lies between application and operating system. It is also defined as a software glue [27]. It permits data management and communication between distributed applications. If two applications are required to be connected for data and database communication, then middleware is used. All the services offered by the operating system are provided to outside applications using middleware. Data management, messaging, authentication, application services are also handled by the middleware. In short, it is connective material to data, applications and users.

The advancement in next-generation computing like AR should be more intelligent and should focus on user needs. To implement such intelligent systems, there is a requirement of special infrastructure as compared with ordinary systems. The system should be structured and flexible and should include fast features and quality mechanisms. In a homogenous or heterogeneous environment, special information is gathered and analyzed it to identify the state of AR system. To offer user-based services, it is important to understand the state of the system. In this case, the key feature is context awareness. Context-aware system (smartphone) can take the benefit of context awareness to offer user-based services as data and applications are context specific [28]. Here, special infrastructure and context cycle operations are required to obtain context information. As the environment is dynamic, the infrastructure

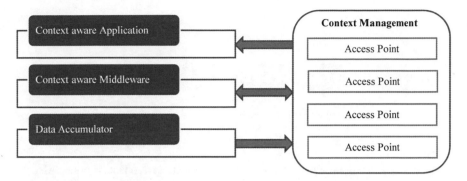

Fig. 3.6 Context-aware system

should deal with run-time data along with other components like toolkits, middleware, libraries, frameworks, architecture, etc. Compared with other components, middleware is vital in developing context-aware pervasive system and providing desired services seamlessly. Middleware is one of the conjoint infrastructure to implement context-aware services by providing context operations like acquisition, access, modeling and delivery in an abstract way.

A middleware includes many different components that permit several functionalities and additional capabilities as per the requirements of a system. To make more efficient context-aware systems, a middleware should have important features related to context-like acquisition, modeling, reasoning and dissemination. Before starting the discussion on middleware, let us consider context-aware system represented in Fig. 3.6.

On the right of Fig. 3.6, the crucial place is context management as all data communication and exchange happens here [29]. All the gathered data from the data accumulator is given to access point in context management using wireless communication network. On the same line, the data gathered can be sent to middleware that can enable desired services. Once data is gathered by the access point from middleware, then it is being sent to context-aware applications. The number of access points is based on the application and its scalability. However, it is ensuring to have the right distribution of access points.

On the left side of Fig. 3.6, there are three components/module-like data accumulator, context-aware middleware and context-aware application. In data accumulator, the smart devices are included like (sensors, smartphones, etc.). In this module, raw data is collected from smart devices. The different data collected can be location information, signal information, etc., Application module is an application that runs on mobile devices. This application has a context-aware feature. In this module, context information is considered, processed, and new information is shown to the user through the application interface. Middleware and its services are represented in Fig. 3.7.

A middleware presented in Fig. 3.7 consists of several phases starting from interface to context distribution. Each phase is discussed here in detail.

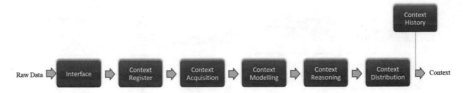

Fig. 3.7 Middleware of context-aware computing system

- **Interface**:
  It acts as entry point where the sensor data collected from access point. Also, the interface sends data to the access point. Once the data is received from access point, it is sent to middleware having sensor ID.
- **Context Register**:
  Here, the context register is used to combine user name and sensor values together and registered with this service. To determine the sensor name, type and user name, sensor ID and user name are matched. Once the sensor ID and name are matched, the corresponding operation is executed. For example, location information generated by the location sensor is sent to locate the engine for its estimation. This is being done to avoid any conflict. This register is updated as soon as there is a new context. Considering a change of user and user environment, there are changes in context information.
- **Context History**:
  It is important module that holds users' context information for future predictions. This storage/database is optimized as it stores only required context attributes. This change of context information is keeping in line with the respective application. This could help to estimate and generate valuable information about users in a specific time interval. For example, the user's name, concerned activity and time spent is stored in the database to identify users monthly working time. The retention of context information is completely depends on the application that it needs or the purpose is fulfilled.
- **Context Acquisition**:
  In the acquisition phase, the raw data received from several input sources [30]. The received data are being processed to generate its value (binary format to reduce overload in communication). In this phase of acquisition, the context tuple attributes are prioritized. A format of tuple includes actuality, location and action. The processing of each attribute is done to decide the priority, and it is then sent to context modeling phase. Based on change in user location or action or actuality, the priority-based actions are performed. For example, if user action are high prioritized compared to location, then change in location information is stored in a data structure (stack) as it is least prioritized. The actuality information is considered to be a change in the user in a domain. There could be two status of user, i.e., either "out of state" or "in state." In short, the highest priority value will go to context modeling else it will be stored in the database and other value calculation begins.

- **Context Modeling**:
  There are several techniques of context modeling [31, 32]. In context modeling, the context values are received from the acquisition phase along with its priority. In this phase, context is represented to understand in the better and efficient way. One of the ways is to use binary tree representation. In this representation, based on the context data like actuality, location and action, the tree is constructed. If the location value is 1, then right child is created, otherwise left child is created. This is applicable for each value of each tuple discussed in context acquisition phase. Based on the tree calculation, the binary streams are generated.
- **Context Reasoning**:
  Context reasoning is also called as decision models. Here, the context is understood in a better way and inferring of new knowledge happens. There are several techniques which exist for context reasoning [33]. One of the techniques is based on logic rules. In this technique, the rules are defined based on a scenario. Considering the tuple attributes and their combinations, meaningful words are created. The words are created according to the rules set in a scenario. For instance, the rules could be the user is in the office or out of office, the user is working or not working, etc. Also, the combination of tuple could be the user is in the office and working. Based on the matching of the rule, the situation is decided, and it is sent to the next phase called context distribution.
- **Context Distribution**:
  The decision made in the context reasoning phase is received in context distribution. After receiving this information, it is transferred to access points in context management discussion in this section. Further, the access point sends this information to a smartphone application using wireless communication networks.

3.3 Context-Aware Architecture

As discussed in the previous section of this chapter, the important feature of pervasive computing or AR is context and context awareness. The visualization and implementation of context-aware pervasive computing or AR ranges from several applications that are running on various platforms like desktop, mobile, handheld devices, etc. To enhance current systems and make smarter system, there is a requirement to characterize the context-aware systems. Interestingly, the characteristics of context information changes as the application changes or system changes or scenario changes. To inculcate the properties of context and context awareness in AR, there is a requirement to revisit the architecture. The context-aware architecture should consider many things. The design and development of context-aware architecture should understand the smart devices integrating with the system, along with each scenario. Also, valuable entities and their observations must be considered in designing the architecture. Addition to this, there are various requirements in designing the architecture of context-aware system. The enriched architecture of the context-aware system empowers technologies to improve the comfort level of every user. The ease in the user's life and the performance of the context-aware pervasive system will be enhanced if the architecture includes all functionality, and it is managed at several levels (layered architecture). The layered architecture will be suitable for large systems where the task composition and execution will be independent.

Let us consider the minimum requirements for designing context-aware architecture like abstraction, embracing resource-constrained devices, scalability and security.

- **Abstraction**:
  In this requirement, the architecture should be capable to hide internal or low-level detailing that handles sensing of information. The data is collected from heterogeneous sensors places in different networks. The designers should ensure that different ways to collect or accept data coming from data sources. Also, the architecture should have the ability to read data continuously or occasionally. The software API should be able to get the data from internal or external sources. The contextual data are multivariate, large in volume and generally updated continuously. Hence, the architecture should manage the complexity of data and accordingly the abstraction methods must be designed.
- **Embracing Smart Devices**:
  Pervasive system is a combination of several components. However, large variety of devices and software components are vital to include. However, few devices are constrained devices in terms of battery power, computational capabilities, etc. On the opposite side, the technology is emerging to support such devices having limitations like limited memory, less computing power and sensing accuracy. The inclusion of such technology advancement is possible and helps to understand context information with modification in the existing architecture. To state one example is an advancement in middleware. As stated previously, the raw data is received from sensor devices and send to middleware for further processing. At middleware, sensed data is received from multiple devices; hence, the middleware processes the aggregated data and can take local decisions. Here, the middleware performs multiple tasks that require additional resources. To overcome the additional task of middleware, the knowledge sharing is vital that share the knowledge across multiple hardware and software components of the architecture. This process will take required resources, and more meaningful information will be produced. This process of knowledge sharing will help the systems to integrate context information and will be the easiest way to develop smart applications [34].
- **Scalability**:
  One of the important observations of the pervasive system is that it has several devices, and devices are of different type. A device could be sensing device, computing device, etc. The devices are constrained devices; hence, there could be a situation where devices work or does not work. Also, the devices should be added or removed from the system as and when it the application demands for. The system may change over a time where there is a change in the functionality

(add, remove, etc.). Hence, the architecture should be layered so that the addition or removal of component is possible with ease. To address scalability, the architecture should support vertical as well as horizontal where the capacity of a system (increasing CPU power, inclusion of additional components, etc.) will be increased. In real-time cases, the system should have functionalities to work the components in sequence or in a parallel way. Hence, the challenge for the designer to decide the architecture can support the working of components based on the requirement. In short, parallelization and concurrency are the vital factors that decide context-aware architecture in a real-time scenario.

- **Security**:
  As the pervasive system provides smart services to users' devices. These services are accessible from any device, at any time. Also, the pervasive system reduces the human intervention to a large scale. There are many services that read user' location as context information. Hence, a lot of personal, location and devise information is collected in a pervasive system. This brings security and privacy issues in front. Therefore, the architecture should include security and privacy measures. The security and privacy principles should be included at the design stage of application and not an add-on type of services. In short, the traditional security and privacy protection mechanism would not work as the devices are frequently changing, the ad hoc network also changes, etc. Hence, the security spectrum should be large enough, and it should include authentication, access control mechanism, privacy, trust management and so on.

A reference architecture of the context-aware pervasive system is depicted in Fig. 3.8. Figure 3.8 consists of many layers starting from physical layer to the application layer. The physical layer is enclosing several devices like sensors, actuators and other devices. There has been continuous invention of these devices. In an architecture, the sensor devices read the environmental data and transfer to the next layer. There are different types of sensors like physical sensors, logical sensors, etc. The physical sensors are camera, microphone, humidity, temperature, GPS, etc. On the other side, equally useful sensors are logical. For example, 1000 online ticket booking, exploring electronic calendar, utilization of healthcare devices, etc.

Once the data are collected by and raw data is being processed and aggregated, then it is forwarded to middleware for further processing. Having said before, middleware is a glue between services offered by the operating system and user applications. The task of middleware is to perform several contexts based operations like acquisition of context, context modeling and reasoning of context and distribution of context. The important task of middleware is context management, device management, adoption of software and hardware agents, discovery of context and services, etc.

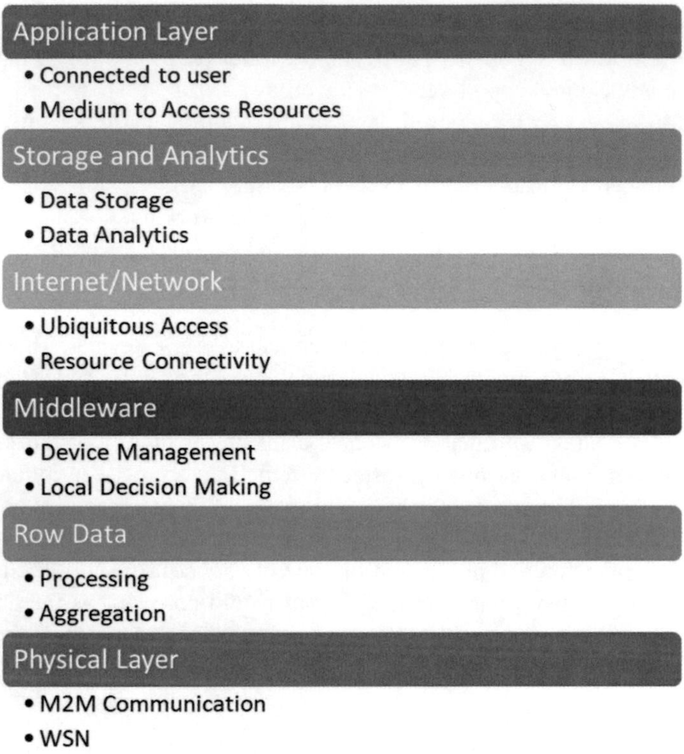

Application Layer
• Connected to user
• Medium to Access Resources

Storage and Analytics
• Data Storage
• Data Analytics

Internet/Network
• Ubiquitous Access
• Resource Connectivity

Middleware
• Device Management
• Local Decision Making

Row Data
• Processing
• Aggregation

Physical Layer
• M2M Communication
• WSN

Fig. 3.8 Abstract layered architecture

3.4   Conclusions

Context and context management are the key features in pervasive computing. The entities communicating with each other reveal valuable information, and this information becomes vital to design AR applications. The applications require to consider entities as well as surrounding environment which could be used, physical or computing environment. This chapter presented context definitions and various types. This chapter also presented a classification of context in detail.

Context and context management shows the key elements in dealing with raw data received from sensing devices. This chapter brings up the importance of middleware and its functionalities in context-aware AR systems. The chapter has discussed in the process of reading raw data and how the raw data is processed to get valuable information shown on user applications.

The chapter also deals with context-aware architecture that focuses on the new requirements compared to architectures of traditional applications or AR applications. The context-aware architecture should be scalable, adoption of all devices and should be secure in the view of open, flexible environment of pervasive or AR system.

References

1. Weiser M (1991) The computer for the 21st century. Scientific American, Sept 1991
2. Feiner S (1994) Redefining the user interface: augmented reality. ACM SIGGRAPH 1994, Course Notes, vol 2, pp 1–18
3. Langlotz T, Grubert J, Grasset R (2013) Augmented reality browsers: essential products or only gadgets? Commun ACM 56:34–36
4. Khriyenko O, Terziyan V (2005) Context description framework for the semantic web. In: Proceedings context 2005 context representation and reasoning workshop. Paris, France. http://sra.itc.it/events/crr05/45.pdf
5. Mahalle P, Dhotre P (2019) Context-aware pervasive systems and applications, https://www.springer.com/gp/book/9789813299511
6. http://www.webster-dictionary.net/definition/context
7. Schilit B, Theimer M (1994) Disseminating active map information to mobile hosts. IEEE Network 8(5):22–32
8. Brown PJ (1996) The Stick-e document: a framework for creating context-aware applications. In: Proceedings of the electronic publishing '96. IFIP, Laxenburg, Austria, pp 259–272
9. Dey AK (1998) Context-aware computing: the CyberDesk project. In: Proceedings of the AAAI 1998 spring symposium on intelligent environments. AAAI Press, Menlo Park, pp 51–54
10. Franklin D, Flaschbart J (1998) All gadget and no representation makes jack a dull environment. In: Proceedings of the AAAI 1998 spring symposium on intelligent environments. AAAI Press, Menlo Park, pp 155 160
11. Harter A, Hopper A, Steggles P, Ward A, Webster P (1999) The anatomy of a context aware application. In: Proceedings of the 5th annual ACM/IEEE international conference on mobile computing and networking (Mobicom'99). ACM Press, New York, pp 59–68
12. Rodden T, Cheverst K, Davies K, Dix A (1998) Exploiting context in HCI design for mobile systems. In: Proceedings of the workshop on human computer interaction with mobile devices, Glasgow, Scotland
13. Hull R, Neaves P, Bedford-Roberts J (1997) Towards situated computing. In: Proceedings of the 1st international symposium on wearable computers (ISWC'97). IEEE, Los Alamitos, pp 146–153
14. Schilit B, Adams N, Want R (1994) Context-aware computing applications. In: Proceedings of the 1st international workshop on mobile computing systems and applications. IEEE, Los Alamitos, pp 85–90
15. Perera C, Zaslavsky A, Christen P, Georgakopoulos D (2014) Context aware computing for the Internet of Things: a survey. IEEE Commun Surv Tutor 16(1):414–454
16. Ranganathan A, Al-Muhtadi J, Campbell RH (2004) Reasoning about uncertain contexts in pervasive computing environments. In: Pervasive computing. IEEE, Apr–June 2004
17. Henricksen K et al (2003) Generating context management infrastructure from high-level context models. MDM 2003, Jan 2003
18. Guelfi N, Savidis A (2006) Rapid integration of software engineering techniques. Springer, Berlin, p 131. ISBN 3-540-34063-7
19. Trullemans S, Van Holsbeeke L, Signer B (2017) The context modelling toolkit: a unified multi-layered context modelling approach. In: Proceedings of the ACM on human-computer interaction (PACMHCI), vol 1, no 1. ACM, pp 7:1–7:16
20. Wang RY, Kon HB, Madnick SE (1993) Data quality requirements analysis and modeling. In: The ninth international conference on data engineering, Vienna, Austria, Apr 1993
21. Dobslaw, Larsson A, Kanter T, Walters J (2010) An object-oriented model in support of context-aware mobile applications. In: Proceedings of 3rd international ICST conference on MOBILe Wireless MiddleWARE, operating systems, and applications (Mobilware '10), ser. LNICST, vol 48, Chicago, USA, June 30–July 2, 2010, pp 205–220
22. Zimmermann Andreas, Specht Marcus, Lorenz Andreas (2005) Personalization and context management. User Model User-Adapt Interact 15:275–302. https://doi.org/10.1007/s11257-005-1092-2

23. Salber D, Dey AK, Abowd GD (1999) The context toolkit: aiding the development of context-enabled applications. In: Proceedings of the 1999 conference on human factors in computing systems, Pittsburgh, PA, pp 434–441
24. Schilit BN (1995) System architecture for context-aware mobile computing. Ph.D. thesis, Columbia University
25. Dey AK, Abowd GD, Salber D (2001) A conceptual framework and a toolkit for supporting the rapid prototyping of context-aware applications. Human-Comput Interact J 16(2–4):97–166
26. Eckel G (2001) LISTEN—augmenting everyday environments with interactive soundscapes. In: Proceedings of the I3 spring days workshop moving between the physical and the digital: exploring and developing new forms of mixed reality user experience, Porto, Portugal
27. What is Middleware? Middleware.org. Defining Technology, 2008. Archived from the original on 29 June 2012. Retrieved 2013-08-11
28. Henricksen K (2003) A framework for context-aware pervasive computing applications. In: Computer Science, School of Information Technology and Electrical Engineering, The University of Queensland, Sept 2003. http://henricksen.id.au/publications/phd-thesis.pdf
29. Bernardos A, Tarrio P, Casar J (2008) A data fusion framework for context-aware mobile services. In: IEEE international conference on multisensor fusion and integration for intelligent systems, 2008. MFI 2008, Aug 2008, pp 606–613. https://doi.org/10.1109/MFI.2008.4648011
30. Chen H, Finin T, Joshi A, Kagal L, Perich F, Chakraborty D (2004) Intelligent agents meet the semantic web in smart spaces. IEE Internet Comput 8(6):69–79. https://doi.org/10.1109/MIC.2004.66
31. Balavalad K, Manvi S, Sutagundar A (2009) Context aware computing in wireless sensor networks. In: International conference on advances in recent technologies in communication and computing, 2009. ARTCom '09, Oct 2009, pp 514–516. https://doi.org/10.1109/ARTCom.2009.85
32. Baldauf M, Dustdar S, Rosenberg F (2007) A survey on context aware systems. Int J Ad Hoc Ubiquitous Comput 2(4):263–277. https://doi.org/10.1504/IJAHUC.2007.014070
33. Bikakis A, Patkos T, Antoniou G, Plexousaki D (2008) A survey of semantics-based approaches for context reasoning in ambient intelligence. In: MM FA, A E (eds) Ambient intelligence 2007 workshops, vol 11. Springer, Berlin
34. Chen H (2004) An intelligent broker architecture for pervasive context—aware systems

Chapter 4
Augmented Reality and IoT

4.1 Augmented Reality

The technology is taking a brisk pace that the things which were not possible yesterday are possible now. Looking at the recent developments like virtual reality (VR) and augmented reality (AR), it is possible to develop several applications that deal with virtual space and physical space. The physical space around the user is a three dimensional (3D). However, traditional devices can show two-dimensional (2D) images. This limitation of understanding real-world objects and their real perception was addressed by Ivan Sutherland in 1960 by inventing a head-mounted display (HMD) [1]. As the user moves, HMD allows the user to see the change in images accordingly. If 2D images are placed on the user's retinas, then it creates an illusion of what the user sees (3D images). The important aspect here is when the user moves his head, then the change should be seen in the image also. In short, the change in 3D images are the same as a change in the user's perspective. In 1990, Tom Caudell defined a new term "AR" [2, 3]. AR is a visualization technology that allows visual models to be played at real locations in real time. The promising fields of AR include building and construction.

There is confusion among VR and AR in terms of their concept, use, etc. Looking at virtual reality (VR), it creates a synthetic or unreal environment. On the other side, AR considers the real environment and superimposes computer-generated objects in that environment [4]. Experiencing the environment in VR and AR is also a point of discussion. In VR, the users believe that they are immersed in simulated or computer-generated reality using headsets. Whereas in AR, the users believe that they are in the real world, it overlays digital content on the real space or environment using smartphones, heads up devices, AR glasses. In short, VR allows you to substitute a real space with virtual space. AR allows you to add virtual information to the real space. The difference between AR and VR is presented in Table 4.1.

AR consists of interfaces that allow interaction between the digital contents of users and their physical environment. These interfaces overlay digital information

Table 4.1 Difference between AR and VR

Augmented reality (AR)	Virtual reality (VR)
AR consist of real as well as virtual worlds	VR generates a virtual world
It helps users to connect with real as well as virtual worlds. It permits the user to differentiate between these two worlds	It allows the user to connect with the virtual world. It is difficult for the user to distinguish real as well as virtual worlds
AR is experienced by using a laptop, smartphone or tablet	VR is experienced by using VR headsets
AR aims to enrich the experience by adding virtual elements (digital image, graphics) as an interaction layer with the real world	VR aims to produce own reality (computer generated)
User stays in the real world	The user is moved from real world to the virtual world
It immerses similar information to existing real-world view	It includes heavy graphics to generate the virtual world
The user experiences a sense of presence in a real world	User experience a sense of presence in a virtual world which is under the control of the system
The user is present at the location of AR experience	The user is not present at the location of VR experience
Users can perform actions on objects (move, rotate) in the real environment	The user cannot perform actions on objects in the virtual environment

(graphics, 3D virtual objects) into the user's surrounding environment in real time. AR allows users to see the surrounding world in real time, made up of virtual objects. Embedding these virtual objects in the user's world is possible using wearable devices. For many years, AR was limited to academic laboratory use cases or sci-fi movies. In the early years, AR systems were used for specific tasks and on an experimental basis. The specific task includes maintenance and repair [5]. Now, AR has reached out to a wide range of applications and its motive varies from case to case. The aim is to provide additional and valuable information to a user that he cannot get if from his senses. The virtual objects of AR give more information that the user is failing to detect the same information using his senses. Hence, AR has a potential capability to use it in several applications in domains like medical, military, advanced driver assistance, entertainment, education, advertisement, smart cities, tourism, etc. [6].

Consider an example of education, where AR has a huge impact in this field. AR brings a constructive change in understanding and learning the things in the future. The survey [7] stated that the understanding of the lesson topic is increased among the students with the help of AR. Also, AR has brought motivation among students and involved them in learning lessons in detail. Using AR, additional 3D information is becoming a powerful and vital tool in the life of the user. This tool is increasing the users' ability to sensing and effectiveness. This will have a huge impact on users' learning, traveling, dealing with diseases, etc.

Particularly, AR caught more attention from users when Google Glass was introduced in 2013 [8]. This has ignited many techies and certain industries to develop and deploy AR systems. For example, Microsoft has released the head-mounted AR device called "HoloLens" [9]; AR plan was developed by Facebook [10], etc. Figure 4.1 shows the head-mounted AR devices (HMARD).

AR systems are not limited to specific wearable devices, but it has been raised to a new level on smartphones. The use of AR on smartphones found to be exponential due to enhancements in AR algorithms [13]. Particularly, in 2016, Pokémon Go was launched by Niantic and Nintendo and it has been downloaded by a million users just in one week [14]. The trend of AR use has been seen clearly in 2017 when Snapchat has released World Lenses [15]. Such apps have been popular and have thousands of active users across the globe. Figure 4.2 represents two popular apps that use AR (Pokémon Go and Snapchat).

This exponential rise in AR apps is due to the following developments observed in recent years.

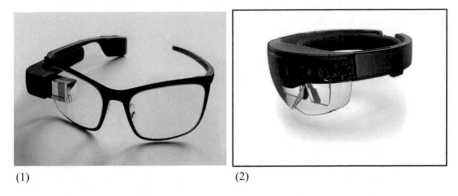

(1) (2)

Fig. 4.1 Examples of HMARD: (1) Google Glass [11] (2) Microsoft HoloLens [12]

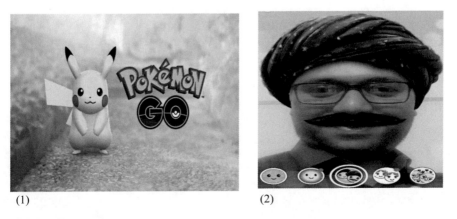

(1) (2)

Fig. 4.2 Mobile AR applications: (1) Pokémon Go (2) Snapchat AR

- **AR and Internet of Things (IoT)**
  AR includes several smart and visual devices. To support this, IoT is playing a key role. As IoT includes several low cost and effective devices in an environment, this makes the applications to be more context-based accurate in performance. Hence, the foundation for AR are the development and availability of smart, tiny and visual devices.
- **Merging of Information**
  One of the important aspects in AR is understanding of the environment and related information. The key idea is to merge/map visual and real information in AR. To do this, the progress in the development of algorithms that perception of the environment (mapping and localization) are essential and becoming vital in AR applications.
- **Enhancement in Optics**
  The scientific study of light and sight has led to the development of AR displays. The display technologies are enriched to support magnification, collimation, and relaying. Hence, AR systems can support the diverse field of user applications.
- **The Readiness of Multimedia Techniques**
  AR applications need to present the content in an improved way. AR applications should be more stylish. Hence, the improvement in multimedia techniques are required.

The advancement in hardware, as well as software, is creating a chance for industries to design, develop and enhance AR applications. Now let us discuss hardware, software components of AR.

- **Hardware Components of AR**

Discussing at the beginning of this chapter, wearable devices are used in AR for user interaction purposes. Most of the hardware components are IoT devices. Each device component is discussed here now.

- **Input Devices**
  These devices help users to interact with AR systems. The interface included in AR acts as an interaction medium between AR systems and concern devices, mostly the important interfaces [16]. For example, the interaction between user and real elements/objects is obtained using tangible interfaces. One of the examples of a tangible interface was used in the VOMAR app [17]. In this application, the user has the choice to reorganize home furniture. The commands are taken using the user's gestures and furniture is rearranged. Another example of an input device is gloves having inbuilt sensors [18]. These gloves would be input to the AR systems (playing games, virtual drawing, etc.). Collaborative interfaces are also important in the field where different medical practitioners can interact with each using video conferencing. They can discuss treatment and line of actions required for the same patient [19]. Another type of input device deals with multimedia interfaces where more than one input a user can use and interact with the AR

system. The inputs can user gesture, blink, touch, speech, etc. These interfaces are becoming more useful as they are easy and robust [20].

- **Sensors**
  These devices are useful in the process of tracking mechanisms. The position of user or device is determined which is essential for visual registration of the physical environment and its digital information. This leads to a merging of digital and physical world images. This is an essential component of AR systems. The scene composition is possible using the tracked data (camera images, 3D models, etc.). The tracking devices could be GPS, ultrasonic, mechanical, etc. These devices are having different resolutions, setup, ranges, etc. which will help to improve the accuracy of AR.

- **Displays**
  These devices allow users to have immediate access to the AR system. These devices like HMD, monitors, wearable devices, handheld devices, optic, etc. allow users to see a virtual environment. HMD includes camera(s) and it is kept on the user's forehead used during gaming, medical application, or even engineering applications. An important use is gesture and voice recognition. This device includes different techniques like holographic, diffraction optics, reflective optics, etc. Another display is the Head-Up Display (HUD) which is compact and lightweight. This device gives additional and valuable information used to focus on the current task. It is used in Google Glass which has a keypad, camera, etc. Using this device, users can take pictures. This device gives information like news, current weather, etc. The next advanced device is HoloLens from Microsoft that permit users to project and interact with colorful objects. This device includes a camera, sensors, microphones, etc.
  Even smart mobile phones are also used for AR applications. As mobile devices are small and act as mini computers, they are easy to interact with AR applications. These devices have big screens, high-resolution cameras, sensors, GPS and powered processors. Using mobile AR applications, the virtual images can be superimposed and made it available on the mobile display.
  Apart from the above-mentioned devices, AR uses processors that are capable to understand visual space and responds as per the AR system.

- **Software Components of AR**

As we mentioned in this chapter, AR varies from application to application. Hence, the software used in the AR application is application-specific. Based on camera and tracking devices, the real-world coordinates are used by the software. For example, location information is captured by creating an XML using AR Markup Language (ARML) [21]. The ARML is represented in Fig. 4.3.

Functional blocks of ARML allow the merging of real and virtual worlds and specify the linkage between these two worlds. This allows for the embedding of virtual objects in the physical or the real world. Based on the user's behavior, these objects are controlled.

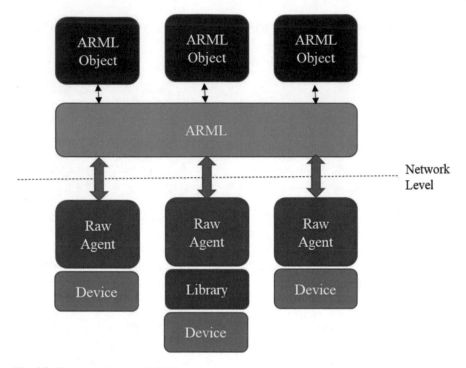

Fig. 4.3 Functional block of ARML

4.2   Augmented Reality and IoT Design Issues

Considering a major revolution in the Internet of Things (IoT) and augmented reality (AR), a considerable change has been observed in the user's life. In this world of communication, everything is connected due to IoT which has received a lot of attention [22]. Alongside IoT, another powerful tool is AR which is used to visualize contextual information about the user, objects and surrounding environment [23]. These two promising technologies are giving an endowment of contextual information next-generation computing. A lot of research has focused on not only technologies but also contextual information. Chapter 3 has already discussed context and context management. Let us consider an AR and IoT-based use case that focuses on context-awareness.

Mr. Harshit is being thirty-eight years old website designer wish to purchase fabrics for him. So, he decides to go to the nearby shopping mall. At the entrance of the mall, his mobile phone notices that he is in a shopping mall. This information is notified of his smart glasses that help him to buy the best fabrics for Mr. Harshit. As there are several options to buy fabrics, Mr. Harshit is confused and not able to decide the best fabrics for him. To solve this problem, the smart glasses start communicating with fabric tags to know the details about each fabric. The details include what is size and color, what is the material is being used, who is the manufacturing brand, what is the price and relevant offers, etc. Once the answers are received, the smart glasses communicate it with Mr. Harshit's mobile to know what Mr. Harshit like, fabrics purchase, size, etc. As Mr. Harshit is a website designer, he prefers to wear casual fabrics rather than professional fabrics. Based on Mr. Harshit's likes and preferences, the phone starts communication with fabrics to get the best suitable fabrics. The responses are shared with smart glass and then the glass highlights the suitable fabrics for Mr. Harshit. The glasses also provide reviews of the fabrics, and reviews from friends who are on Facebook, etc. To improve better choice, the summarization of relevant fabrics is listed on the smart glass that helps Mr. Harshit to make the final choice.

The above scenario states the vision of AR and IoT keeping context-awareness in consideration. This scenario explains how IoT, AR and context management are involved and makes the user's task easy.

Even though there are several efforts put in by researchers on AR and IoT, there are a few practical issues [24]. Each issue is discussed here now.

- **Interoperability**
  In IoT, the things/devices should communicate with each other seamlessly and must produce the desired result for users. In this case, these devices are different in several aspects like their size, configuration, purpose, etc. Hence, the important issue in designing an AR-based IoT solution is interoperability [25]. Therefore, the solution should understand the semantic communication among these devices. This means the solution should know the devices, their purposes, etc. which could lead to understanding the semantics of objects. These objects also should be in the same interest. Having the idea of each object, they can collaborate among themselves and work for a common goal. Hence, the necessary technology in AR-based IoT solutions should consider the communication between unknown objects. Leveraging the semantic web would be useful to enrich the digital content and could be displayed in the user interface of AR.
- **Security and Trust**
  The next issue is security as the things or devices will communicate in an AR-based IoT solution. The devices are known to each other in some case; however, a device is unknown to other devices. Therefore, the communication between such devices should be secured and should build trust over frequent communication. The solution should check the authenticity of each object, secure communication and guaranteed delivery of information. Before any untrusted device or virus ruin the AR environment, the necessary security measures should be considered in the design phase of AR solutions.
- **Context-Aware**
  The communication between trusted things shall communicate flawlessly and exchange of information should be carried out. The information exchanged among things should be real time, up-to-date and relevant to the context of user requests. The issue is keeping a track of information which is context centric or the applications should be context-aware [26]. This will help the system to avoid information conflict across many smart devices in a network. The challenge is to maintain consistency and context-aware is information available to resource-constrained smart devices in the AR-based IoT solutions.

- **Minimum User Intervention**
  Looking at the AR and applications, an issue is the dependency on the user and user intervention. Autonomy should be given to each thing in an IoT environment that makes everything invisible to the user's perspective. Each device or thing should be self-dependant, self-learner and must be reactive in the communication. This will lead to developing an autonomous ecosystem.
  One of the important failures is a visual sensor accelerometer. There are different reasons; however, an important reason is the rendering of 3D images on smart, but resource-constrained devices like phones. As cell phone camera captures 2D images, the 3D rendering becomes heavy. The issue is how to improvise camera performance and GPS accuracy. Another issue in the hardware part is the heaviness of the device due to its size. However, AR leading companies are addressing this issue by developing compact and low-weight devices.
- **Hardware Issues**
  The requirement of designing the devices in IoT is less energy consumption. The range of devices used is ranging from less to more powerful processors (like 8 bit to 32 bit). On the same line, these devices should work in different environments and platforms (like Atmel, Cortext, and different device interfaces). Zigbee [27] is one of the hardware platforms that support different wireless interfaces in small, medium ranges. On the same line, the open-source platform includes Raspberry Pi [28], Arduino [29], and BeagleBone [30]. One of the important failures is a visual sensor accelerometer. There are different reasons; however, an important reason is the rendering of 3D images on smart, but resource-constrained devices like phones. As cell phone camera captures 2D images, the 3D rendering becomes heavy. The issue is how to improvise camera performance and GPS accuracy. Another issue in the hardware part is the heaviness of the device due to its size. However, AR leading companies are addressing this issue by developing compact and low-weight devices.
- **Software Issues**
  As there are different hardware platforms, the issue is the development of software compatibility with these hardware platforms. As AR applications are running on from desktop to mobile and small tiny devices, there are issues in the functioning of software on a hardware platform. As the devices are changing regularly at large scales, software interoperability is a big issue. As the devices are resource-constrained (limited memory and computing power), the operating system (OS) operations should below. OS of IoT should support the heterogeneity of devices and must be optimized in code, size, and power. The examples of OS are TinyOS [31], FreeRTOS [32], and OpenWSN [33]. The TinyOS is an open-source OS, which is embedded in nature. FreeRTOS support multiple architectures and written in C language.
  The specific AR browsers are working, but the difficulty in interacting with social media. Another issue is the scalability of applications in an AR system. There is a requirement to develop a set of toolkit that could support several devise and

applications running across multiple platforms. A single interface could act as multiple interfaces wherever possible.

4.3 Augmented Reality-Based IoT Use Cases

Using AR, digital contents are superimposed and intermixed into the user's sight and the real world is formed. The tasks performed by AR consist of capturing and identification of the scene. This will help to get accurate information. Once it is done, the further task includes processing of the scene and its visualization. These common steps are executed on several devices and help to create many use cases [34]. In this section, different use cases are discussed.

- **Use Case: Medical**

Innovation and invention are aiming to improve the user's life. AR and IoT can improve the healthcare industry. Due to wireless sensor networks (WSNs), the healthcare sector has changed radically. Not only WSN but wireless body area networks (WBANs) have brought a medical application that has made significant changes in the healthcare domain. IoT-based medical applications are aimed to provide special features like remote monitoring of patients, identification, and preventions of critical patients, support to senior patients using smart environment, etc. [35]. Also, monitoring and handling medical databases, context-based patient services are provided.

Safety and efficiency are the major two features of new technology. The significant changes made by AR in the medical field include diagnosis, education to interns (future doctors), surgery, etc. AR is also helpful in identifying depressions or stress by reading facial expressions, physical gestures, voice tones, etc. [36].

There are many times it is observed that a patient faces difficulty in telling doctors what exactly he is bothered about. To identify the symptoms of a disease, the AR-based simulators will help the patients as well as doctors to make decisions.

The simulation and visualization of medical services are of high importance due to several reasons. Choosing the correct platform to analyze, visualize and provide supportive services is the need of an hour. In this case, the AR will provide any assistance to medical practitioners. For instance, AR use in the medical field is an enriched view of a fetus in a woman's womb. On the same line, the surgeons can take the help of AR systems to understand, visualize the patient's medical information to identify tumors or lumps. This will help surgeons to make the right decision in a quick time. The head-mounted devices can be used to render the infected body part of a patient.

AR has the potential to enhance the effectiveness and depth of medical training in many ways. For future medical doctors or interns, AR headset is very useful to provide an insight view of the body with an interactive 3D structure. The graphical environment created by AR-based camera which will help the interns to understand

and receive detailed information about body parts. Not only understanding of the human body, but AR can also provide information about diagnosis, surgery, medication and recovery of the patient in a graphical environment using AR devices and applications.

- **Use Case: Classroom Education**

Starting from laptops, tablets, interactive projectors have been widespread in colleges and schools. The students and teachers are ramping up toward the e-learning environment. This ramping up of digital learning is leading to AR-based classroom education.

Using AR applications on a smartphone or tablet, a learner can get a more enriched environment in the classroom. Using the AR application, a student can understand astronomy by visualizing a complete map of a solar system sitting in a classroom. Similarly, a learner can learn a musical instrument by seeing musical notes in a musical classroom.

Also, AR is becoming a part of the standard curriculum nowadays. In this case, several course content materials of different formats (text, audio, video, and graphics) are superimposed with the student's environment. The course content materials are being scanned into the AR system and then it is being rendered in the multimedia format for a better understanding of the technical concept to the learners. Inclusion of 3D models, engaging actions and sound are included in interactive AR books. Also, AR workbooks or worksheets are used to link homework, lesson, classwork, etc.

Field visit or industrial visit is enhanced by providing real experience and more information using AR glasses at visiting industrial places. Historical information is being presented to learners using multimedia techniques. The difficult courses are explained in the easiest way of using AR. Here, the contents of mathematics or engineering are rendered. In subjects like chemistry, the learner can see and interact with the molecular structure using handheld devices in AR applications.

- **Use Case: Marketing**

Purchasing a product is a part and parcel of the user's life. Whenever a user wishes to buy a product, normally he does a review or analysis of that product. By browsing the Internet, a user can get the information. However, this process is time-consuming. To make the user's life easier, AR applications can help them [37]. For an instance, in a grocery shop, users can use AR glasses and AR-based cameras to get information about the product like name, prices, reviews by earlier buyers, location of the product in that shop, etc.

The use of AR at a large scale is being observed in the field of the automotive industry. Such industries can reach out to customers easily by making 3D modeling of vehicles and then it is used for advertisement. This method of creating visual models using AR has been used in markets for saleable items, like furniture, home, shoes, etc.

To visualize the final output at the time of creating process is the advantage of AR applications. For example, a person wishes to renovate his office or home, and then an AR headset helps persons to enter into the office directly and can help him by giving several ideas using 3D modeling and visualization techniques. The entire layout of the office is made available to the user on his AR device. The internal structuring of office (electricity and plumbing) will reduce user efforts. Also, the interior of the office can be made possible by making the required real changes. The AR applications can render the different interior options for office as well as homes.

For example, a user is asked to wear a pair of socks. Further, an AR camera captures the images of socks and then the user can see the desired shoes for him. The AR systems allow the customers to see the shoes with additional information like size, color, price and other accessories. The use of AR has been prominently observed in computer games and movies.

The AR systems provide an equal opportunity to both customers and manufacturers [38]. This will be a great help to customers to know the products before he buys them. Also, AR helps the manufactures to understand the user and his interests or preferences to provide user-centric services.

- **Use Case: Entertainment**

The entertainment industry is enriched with the use of AR systems as it is a more suitable way to reach out to customers with the desired applications in terms of games, movies, advertisements, etc. Using AR-based entertainment sectors, the user is completely immersed in the real environment of entertainment.

The use of the 360-degree camera can be used to shoot a film. Based on those data, AR applications can render that film where the audience can believe the real experience of movie or film. In games, visibility is an important aspect and it is achieved using AR. The experience of video games is heightened by AR applications. The AR systems include both hardware and software components that create an opportunity for users to submerge into the virtual world. In gaming, AR converts the physical environment into a videogame canvas. One of the popular examples is Pokémon Go, a crazy game downloaded across the globe.

In the AR game, a person has to hit the click button on a video game icon. Once the AR game begins, a person can grab a gun kept on a desk and stats battling with robots who are forcefully entering into a home or office. Every action seems to be real.

The different environments of games like racing tracks, cricket ground, swimming pools, etc. are easily created or prepared. For annotations of different things are possible using AR like trajectories of the ball, car racing, the performance of life swimmers, etc.

- **Use Case: Tourism**

One of the important pillars of the nation's economy is tourism. The tourist attraction is vital and needs enhanced technology to reach out of the tourist. The challenge is to provide context-based services to the tourist when it comes to a large number. AR provides a kind opportunity for agents, tourist brands and travelers to have impressive experience of tourism. This experience can be gained before a person travels.

A "virtual walk" using AR glasses in a favorite city can be a real experience before your book travel trip. A tourist faces difficulty to gain the information of the location before he visits that location. For travelers, AR found to be a platform to visualize and gain information about the different locations. The location information, features, comments or reviews posted by previous visitors are provided by AR applications.

Navigation is another issue for any tourist to reach the locations in a short time. If a tourist is in a new area or city, it is a time burning situation for him and that the AR application supports the tourist by providing navigation and maps. The AR-enabled pointing device can be used by directing it toward a transportation symbol or object. Then, on your device, a tourist can get valuable information like a map of the city, directions from your current location, city bus schedule, etc.

The rendering of historical places, objects, and the events is rendered and a tourist can see it on the landscape. To enhance tourism, unknown places are rendered using AR and attract tourists to such places with more information.

4.4 Augmented Reality and IoT-Enabled Architecture

AR is to blend the digital world and the physical world. In AR, a user has immersed himself in AR experience (mix of the physical and digital world). The great feature of the AR system is that it gives real-time experience. AR systems are not only enhancing user experiences but also providing a business opportunity for service providers. The applications of AR have been seen across multiple sectors or domains. Considering development in mobile phones, sensors, camera quality, tracking technology, and wireless network empowering AR applications to be considered and implemented in a mobile environment. However, one thing to be noted here is that AR systems target on one system or subsystem like human–computer interaction, position tracking, etc.

However, there are certain attributes required to consider in designing the architecture of AR. The quality attributes are classified as a high priority and low priority attributes. Considering the immersing property of AR, the high priority quality attributes include latency in rendering and tracking, network connection, handling multiple devices, reusability and integrating new architecture with the existing one. On the other side, the low priority quality attributes include re-configurability, security, and uptime of the system, fault tolerance, user preferences and multiple hardware support.

Figure 4.4 represents the architecture of AR that includes standard modules that could be generally found in any AR applications.

There are six subsystems of the overall architecture like AR input, AR context, AR tracking, AR application, AR object and AR output. Each subsystem is discussed further. However, it does not necessarily represent a sequence of designing AR architecture.

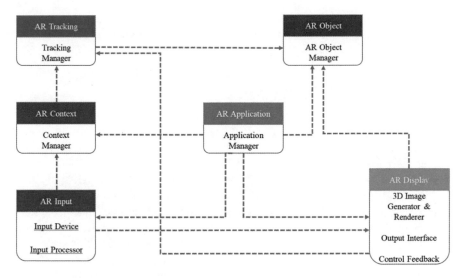

Fig. 4.4 General architecture of AR

- **AR Input**

  In this subsystem, many devices and technologies are included to detect input from the real world and related components. This includes RFID tags, markers, barcodes, cameras, etc.). This module gives information about the position of the user or object and their orientation. For example, it includes location, a gyroscope to have angular velocity, accelerometer, magnetometer and GPS in the smartphone, etc. Along with mentioned devices or sensors, this subsystem includes technologies that allow users to have seamless interaction with AR systems or applications like haptic controller (touch control), camera, etc. Having said earlier, the important element of this subsystem is a sensor as they are responsible to get real information for the execution of AR applications with correct accuracy [39]. The sensors provide valuable information about users and objects like position, location and orientation. Environmental values like lightness, temperature and humidity are received in the AR applications using sensors. Apart from this, this subsystem also deals with sensors in the bodysuit, glove, wand, etc. The technologies required for voice, gesture and gaze detection are also covered in this input subsystem.

- **AR Display**

  User senses and their signals are handled in AR applications by this subsystem. This subsystem includes a display for various services like visual, audio, haptic/touch, sensor, etc. For example, a computer screen, tablets and smartphones are used to provide visual display services in AR applications. Headphones, earphones are a few devices used for audio display services. For sensing touch, gloves are used as display devices [40].

  On the other side, this subsystem includes other technologies for rendering based on the input signal. The main purpose is to provide users a feel of their presence in the real environment. The user's senses are stimulated using these output displays. 3D display, HMD, screen, wall projectors are few names of devices used for visual rendering.

- **AR Context**

  In this module, context and context-related information is collected so that it is available to other modules whenever required. In this subsystem, providing high-quality content is mandatory. Also, the content should be context-aware as the context is an important thing in AR. Creating, transferring and storing content as well as the context is considered using the software in this subsystem (cloud, context management, etc.)

  According to the user's context, the contents are provided to the AR application subsystem. The digital layer in AR application is augmentation which comprises information in audio, video, text and other formats. To create the content, the skills may require or may not. It completely depends upon the AR application to develop. The computer vision, pattern and position tracking, image recognition techniques are required in some cases [41, 42]. The use of a high-quality camera can give facts in a real environment from different angles and places. Providing access to context and content requires sharing services using an Internet connection.

- **AR Tracking**

  Based on the application, this subsystem is used for tracking and management of objects or users. This subsystem control activities of the user, what they see, highlight relevant things. This subsystem is linked to the AR application subsystem. AR applications are controlled using this subsystem. The start, stop and pause of AR applications are performed using monitoring technologies. The controlling and tracking of events and users will result in log files and it is handled by reporting tools.

- **AR Application**

  This subsystem includes content and logic related to a specific application. This subsystem also defined as a processing subsystem that deals with many tasks. Based on the input signals received in the AR input subsystem, the instructions are executed as per the logic. After this, the signals are generated and the demonstration begins using an AR display subsystem. It has many processing units like graphics as well as a microprocessor.

  To generate or create an AR application, this subsystem deals with sensors, rendering units, etc. This subsystem is very crucial as the application will be launched as soon as there is a change in the user location or change in user input. This subsystem includes VR for the creation of immersive content, input data acquisition software and graphics engine. This includes software for sensor integration, software for gesture and image recognition, software for content acquisition, etc.

- **AR Object**

  This subsystem is dealing with all the objects in AR applications. The information about real and virtual objects is stored in this subsystem. This information is provided to other modules whenever they are requested.

4.5 Conclusions

The blend of AR and IoT has proved to be beneficial for many people working in many fields. The immersive experience to users takes AR to a new level in emerging technologies. This chapter has presented an overview of AR. AR technology aims to enhance the user experience by adding an augmentation layer during the interaction with the real world.

Even though AR technology is a promising technology, there are challenges and issues. This chapter also mentioned design issues that need to be addressed to make a well-accepted AR application. The chapter also summarized the AR base IoT use case. AR use is not limited to any particular field as the recent development in IoT and technology is suitable for AR applications.

This chapter also summarizes the architecture of AR. AR is a domain-specific application-specific, and hence, the challenge is to design a general architecture that suits every AR application. This chapter has presented a general architecture which includes required elements or subsystems.

References

1. Sutherland IE (1968) A head-mounted three dimensional display. In: Proceedings of American federation of information processing societies (AFIPS), pp 757–764
2. Lee K (2012) Augmented reality in education and training. Techtrends: Link Res Pract Improve Learn 56(2):13–21
3. Azuma R, Baillot Y, Behringer R, Feiner S, Julier S, Mac-Intyre B (2001) Recent advances in augmented reality. IEEE Comput Graph Appl 21(6):34–47
4. https://www.forbes.com/sites/bernardmarr/2019/07/19/the-important-difference-between-virtual-reality-augmented-reality-and-mixed-reality/#7b74309c35d3
5. Henderson S, Feiner S (2017) Augmented reality for maintenance and repair (ARMAR). http://graphics.cs.columbia.edu/projects/armar
6. Curran K, McFadden D, Devlin R (2011) The role of augmented reality within ambient intelligence. Int J Ambient Comput Intell (IJACI) 3(2):16–34
7. Hsu YS, Lin YH, Yang B (2017) Impact of augmented reality lessons on students' STEM interest. Res Pract Technol Enhanc Learn 12(1):1–14
8. https://www.google.com/glass/tech-specs/
9. www.microsoft.com/en-us/hololens
10. https://adage.com/article/digital/mark-zuckerberg-confident-facebook-lead-snapchat-ar/308724
11. https://en.wikipedia.org/wiki/Google_Glass
12. https://en.wikipedia.org/wiki/Microsoft_HoloLens
13. Casas S, Olanda R, Dey N (2017) Motion cueing algorithms: a review: algorithms, evaluation and tuning. Int J Virtual Augment Real (IJVAR) 1(1):90–106
14. Molina B (2016) 'Pokémon Go' fastest mobile game to 10 M downloads, USA Today, 20 July 2016. www.usatoday.com/story/tech/gaming/2016/07/20/pokemon-go-fastest-mobile-game-10m-downloads/87338366
15. Newton C (2017) Snapchat adds world lenses to further its push into augmented reality, The Verge, 18 Apr 2017. www.theverge.com/2017/4/18/15333130/snapchat-world-lenses-something-new-for-facebook-to-copy

16. Carmigniani J, Furht B, Anisetti M, Ceravolo P, Damiani E, Ivkovic M (2011) Augmented reality technologies, systems and applications. Multimed Tools Appl 51(1):341–377
17. Kato H, Billinghurst M, Poupyrev I, Imamoto K, Tachibana K (2000) Virtual object manipulation on a table-top AR environment. IEEE
18. Van Krevelen D, Poelman R (2007) Augmented reality: technologies, applications and limitations
19. Barakonyi I, Fahmy T, Schmalstieg D (2004) Remote collaboration using augmented reality videoconferencing. In: Proceedings of graphics interface 2004, Canadian human-computer communications society, pp 89–96
20. Lee J-Y, Lee S-H, Park H-M, Lee S-K, Choi J-S, Kwon J-S (2010) Design and implementation of a wearable AR annotation system using gaze interaction. In: International conference on consumer electronics (ICCE), digest of technical papers, pp 185–186
21. Nishimura K (2004) ARW framework and intelligent character agent. IPSJ SIG Tech Rep 2004(28):95–100 (in Japanese)
22. Atzori L, Iera A, Morabito G (2010) The Internet of Things: a survey. Comput Netw 54(15):2787–2805
23. Ajanki A et al (2010) An augmented reality interface to contextual information. J Virtual Real 15:1–13
24. Hazlewood WR, Coyle L (2009) On ambient information systems: challenges of design and evaluation. Int J Ambient Comput Intell (IJACI) 1(2):1–12
25. Park H, Kim B, Ko Y, Lee D (2011) InterX: a service interoperability gateway for heterogeneous smart objects. In: 2011 IEEE international conference on pervasive computing and communications workshops (PERCOM 2011 workshops), Mar 2011, pp 233–238
26. National Intelligence Council, Disruptive Civil Technologies—six technologies with potential impacts on US interests out to 2025 –Conference Report CR 2008–07, Apr 2008, http://www.dni.gov/nic/NIC_home.html
27. Texas Instruments, CC2538 powerful wireless microcontroller system-on-chip for 2.4-GHz IEEE 802.15.4, 6LoWPAN, and ZigBee Applications. Available at: http://www.ti.com/lit/ds/symlink/cc2538.pdf
28. Raspberry Pi platform. Available at: http://www.adafruit.com/pdfs
29. Arduino platform. Available at: https://www.arduino.cc/
30. BeagleBone platform. Available at: http://beagleboard.org/bone
31. TinyOS Operating System. Available at: http://www.tinyos.net/
32. FreeRTOS Operating System. Available at: http://www.freertos.org/
33. Watteyne T, Vilajosana X, Kerkez B, Chraim F, Weekly K, Wang Q, Glaser S, Pister K (2012) OpenWSN: a standards-based low-power wireless development environment. Trans Emerg Tel Tech 23:480493. https://doi.org/10.1002/ett.2558
34. Couderc P, Banâtre M (2003) Ambient computing applications: an experience with the SPREAD approach. In: Proceedings of the 36th annual Hawaii international conference on system sciences, Jan 2003. IEEE, p 9
35. Fang H, Dan X, Shaowu S (2013) On the application of the Internet of Things in the field of medical and health care. In: Green computing and communications (GreenCom), 2013 IEEE and Internet of Things (iThings/CPSCom), IEEE international conference on and IEEE cyber, physical and social computing, 20–23 Aug 2013, pp 2053–2058. https://doi.org/10.1109/greencom-ithings-cpscom.2013.384
36. Sudha MR, Sriraghav K, Jacob SG, Manisha S (2017) Approaches and applications of virtual reality and gesture recognition: a review. Int J Ambient Comput Intell (IJACI) 8(4):1–18
37. Zanella A, Bui N, Castellani A, Vangelista L, Zorzi M (2014) Internet of Things for smart cities. Internet of Things J IEEE 1(1):22–32. https://doi.org/10.1109/JIOT.2014.2306328
38. Biswas A, Dutta S, Dey N, Azar AT (2014) A Kinect-less augmented reality approach to real-time tag-less virtual trial room simulation. Int J Serv Sci Manage Eng Technol (IJSSMET) 5(4):13–28
39. Clements P, Kazman R, Klein M (2002) Evaluating software architectures: methods and case studies. Addison Wesley, Boston

40. Craig AB (2013) Understanding augmented reality: concepts and applications, Newnes
41. Ronak Dipakkumar G, Patel DS (2018) Virtual reality–opportunities and challenges. Virtual Real 5(01)
42. Cubillo J, Martin S, Castro M, Boticki I (2015) Preparing augmented reality learning content should be easy: UNED ARLE—an authoring tool for augmented reality learning environments. Comput Appl Eng Educ 23(5):778–789

Chapter 5
Conclusions

5.1 Summary

This section concludes the book and proposes the future outlook which can be explored further to initiate new research avenues. The objective of this book is to discuss the integration of IoT with AR. IoT is the service network that converges sensor nodes, RFID objects and smart devices. IoT connects objects around us (electronic, electrical, non-electrical) to provide seamless communication and contextual services provided by them. Communication technologies like Wi-Fi, Bluetooth, Zigbee, LoRaWAN are improved on a great scale to be suited for IoT. In the first chapter, IoT building blocks, IoT layered architecture and challenges of IoT are discussed.

IoT makes user life comfortable by providing services in various domains, like smart homes, intelligent transport, smart city, etc. To understand more about IoT and its applications, various use cases of IoT are discussed in Chap. 2. For each scenario, a study is targeted toward the type of IoT devices needed, method of organizing devices, type of communication technology used, services offered, user roles and the mechanism of the service advertisement, and security aspects. Furthermore, moving ahead of IoT, to make life more comfortable, we try to innovate more. We make use of advanced and emerging technology to make our life easier. Here comes the role of AR, how AR can improve these IoT use cases is also discussed in Chap. 2 by providing various possibilities of AR to improve these use cases. When IoT is combined with AR technology, it will provide better visualization as well as real-time information. A most important advantage of combining AR with IoT is to cover the gap between the real and digital world by actual interaction with the real world in real time. AR devices (mounted/portable) collect real-time data from the current environment (context), and according to what the user is currently doing, only information needed for current work is intuitively displayed.

Subsequently, only we can say that context information is an important pillar to realize IoT-enabled AR applications; hence, framework for context retrieval and

G. R. Shinde et al., *Internet of Things Integrated Augmented Reality*,
SpringerBriefs in Computational Intelligence,
https://doi.org/10.1007/978-981-15-6374-4_5

processing is required. Context and context management are the key features in pervasive computing. The entities communicating with each other reveal valuable information and this information becomes important for context-aware applications. Applications require to consider entities as well the as surrounding environment which could be the user's physical or computing environment. In line with this, Chap. 3 presents context, classification of context and scheme for context management. There is a need for architecture to build context-aware applications; requirements of context-aware architecture are also discussed in this chapter [1].

The blend of AR and IoT has proved to be beneficial for many people working in many fields. The immersive experience to users takes AR to a new level in emerging technologies. AR combined with IoT can be used for a wide array of applications like medical, military, advanced driver assistance, entertainment, education, advertisement, smart cities, tourism, etc. AR considers the real environment and superimposes computer-generated objects in that environment. AR consists of interfaces that allow interaction between the digital contents of user and their physical environment. AR allows users to see the surrounding world in real time, made up of virtual objects. Embedding these virtual objects in the user's world is possible using wearable devices. Chapter 4 has presented an overview of AR, hardware and software required for IoT-enabled AR applications. Every technology need to face challenges in its initial phase, there are various design issues of IoT-enabled AR applications, these design issues are also presented in Chap. 4. To understand more about IoT-enabled AR applications, five different use cases, i.e., medical, classroom education, marketing, entertainment and tourism are presented in this chapter. Finally, this book presents the architecture for IoT-enabled AR applications.

The entire summary of this book is depicted in Fig. 5.1; important functionalities of the IoT-AR integration platform are management of data, device, context information, object control and applications. These are building blocks of any IoT-AR integration. The application management includes management of all services that required for discovering IoT-AR services and AR devices in the surrounding. Context management is required for interactive services and this data is collected from various IoT devices; hence, data management also plays an important role. Applications are designed for its basic functionality of controlling objects; hence, it is a core part of IoT-AR. The AR-based application requires a Tangible User Interface (TUI) instead of Graphical User Interface (GUI) where users can interact with digital information using the physical environment. TUI helps to enrich IoT-enabled AR applications with more inter-activeness.

5.2 Open Research Issues

IoT integrated with AR enriches very useful use cases across various verticals starting from education to engineering as stated earlier. IoT and AR have a common set of hardware and software components from the perspective of application development. Hardware components include sensors, RFID, smartphones, and displays

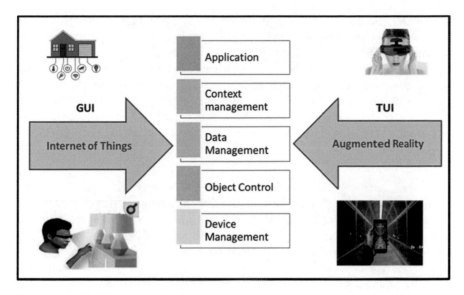

Fig. 5.1 Building blocks in IoT-AR integration

wherein software components include markup languages like ARML and programming languages like Python and Java. Apart from this, other functional components required for IoT-AR use commercial use cases are wireless technologies, protocol and platforms. Wireless technologies include Bluetooth 3.x, 4.x, Zigbee, Wi-Fi, 6LowPAN, Thread 2.5G/3G/4G, LoRA and SigFox which are required to connect AR device to the object under consideration. Protocols play important role in IoT-AR convergence essentially gateway protocols like COAP and MQTT [2] which are required to connect offline use cases to the outside world through the Internet. When the IoT-AR use case is in function, it collects a lot of data from the associated sensors and these gateway protocols play a crucial role in posting this data on the cloud. However, the selection of an appropriate protocol is also another question that needs to be addressed. Selection of suitable platform for the process this AR data is equally important for ease of use and other issues like security and privacy. The use of these platforms is twofold. These cloud platforms provide various services like storage, update and access. Among all services, storage service is the basic feature of these platforms. Storage services provide a mechanism to store and handle heterogeneous and high volume data. The end-user can also access this data through various cloud applications through their IoT-enabled smart devices. Some proprietary popular platforms are provided by IT giants like IBM, Microsoft, Google and Amazon. The open-source includes Thinger.io and iot.eclipse.org and the big data tools required for analytics are Hadoop, HANA, MongoDB, Splunk, Google Charts, etc. Electronic platforms to process sensor data locally are ARMmbed, Arduino, Raspberry Pi, Beagle Bone, etc. The selection of a suitable sensor processing platform depends on the underlying application of IoT-AR.

However, there are many open research areas in integrating IoT with AR. When we look at this integration, there are various challenges and design issues while making spaces more proactive, adaptive, smarter and interactive. It should be noted that IoT and AR have their own set of targeted applications and objectives. However, literature shows that their convergence can enrich the applications for more business values [3, 4]. The main design issues in developing IOT-AR applications are resource constraints of lightweight IoT and AR-enabled devices, distributed services, and operations, data management, etc. These open research issues are discussed below.

- *Scalability*

Due to the large number of devices connected to the Internet, scalability is an important design issue of any IoT application. Consider an example to develop a real-time IoT application like smart home or smart parking which is AR-enabled. Example: Smart home when the user enters into a house the required resources getting on like fan, light by sensing environment means if sunlight is sufficiently available then no need to be switch on a light. If anyone comes at the door, the camera module automatically captures his image and sends it to the email account of the user or sends a notification to the user, and if the user wants to open the door, then he just sends a signal so that with the help of actuator door can be opened. In this application, it is very important to enable AR services for all available devices without losing the performance parameter like latency. It is expected by the end-user that IoT-AR application should be able to support the required number of objects or devices without significant latency. In the sequel, more research is required to make these IoT-AR-enabled applications more scalable without compromising performance metrics like latency, throughput, ease of use, packet delivery ratio and interactive nature.

- *Context Management*

Pervasive AR is an important feature of AR application where latency-aware (i.e., continuous, i.e., zero-delay response) and multi-user and multipurpose user interactions are the key features. For associating such features, context adaptation, its management and making AR applications adaptive to these changing situations are key steps. To design context-aware applications, there is a need for intelligent algorithms for context sensing, classification of context based on the sources and for pervasive augmented reality. The main context sources include humans, the environment and the system [5]. However, modeling of these context sources and integrating them with AR services need further research.

- *Data Management*

All devices connected to IoT-AR application generate a massive amount of data collected from heterogeneous and this data is posted on the cloud. This data is complex and this device-centric management of data along with associated AR services is a bigger challenge. In this case, legacy methods of data analytics do not work efficiently. In the sequel, there is a need to explore more techniques to process this big data to analyze as well as get meaningful insights from this data.

- *Access Control*

The distributed access control mechanism in a decentralized environment where IoT devices are located at multiple geographical locations is a challenging task. As stated earlier, in IoT-AR applications, sensor data is placed at multiple locations and AR-enabled devices require access to this data continuously for providing various services to the users. Designing a scalable and secure access control algorithm for this distributed sensor data requires further study so that the quality of services can be optimized.

- *Security and Privacy*

Many IoT-AR applications are ubiquitous nature where operations and transactions are anything, anytime and anywhere in nature. In such cases, device-to-device communication on the fly through various wireless medium should be secure and confidential. Security by design and privacy by design are emerging approaches that need to be researched further and to be incorporated in designing attack-resistant security algorithms for these applications.

- *Performance*

IoT is the convergence of devices like sensors, smart devices and RFID devices which are all resource-constrained. These resources include limited memory, limited computing power and battery. Minimum human intervention is a key requirement for such IoT-AR applications; most of the processing is carried out on these devices. Therefore, algorithms and mechanisms required for service discovery, device discovery should be lightweight and further research is required to optimize the performance metrics like latency, throughput, packet delivery ration, etc.

5.3 Future Outlook

There are many issues and challenges related to IoT and AR integrations which have not been considered the scope of this book. There are several opportunities and research avenues that can be explored and build on the top of the contents presented in this book. This section presents these possibilities and open issues for future research.

IoT, AR and their integration have resulted in many rich use cases. IoT devices are diverse in size, functionality, the type of data they sense and generate, the format in which data is aggregated. In this vie, IoT-AR application requires standard data formats to support object attributes in 3D fashion. For commercial IoT use case which is AR-enabled, the size of a dataset can be huge and required more processing resources. In such cases, the data is real time and the velocity is another important area that will require real-time data analytics methods. In addition to this, these applications will require better data visualization techniques for clear understanding.

As there is minimum human intervention in IoT-AR applications, resource-constrained IoT devices are always engaged in processing. However, we will require some delegation mechanism for faster and robust detection and tracking of the objects. In addition to this, AR data management, guided tracking of the objects, object control and interface based on AR, generic framework for IoT-AR applications can be also important areas to explore further.

References

1. Mahalle P, Dhotre P (2019) Context-aware pervasive systems and applications, https://www.spr inger.com/gp/book/9789813299511
2. Gündoğan C, Kietzmann P, Lenders M, Petersen H, Schmidt TC, Wählisch M (2018) NDN, CoAP, and MQTT: a comparative measurement study in the IoT. In: Proceedings of ACM ICN, ACM
3. Jo D, Kim GJ (2019) IoT + AR: pervasive and augmented environments for Digi-log shopping experience. Hum-Centric Comput Inf Sci 9:1. https://doi.org/10.1186/s13673-018-0162-5
4. Jo D, Kim GJ (2019) AR enabled IoT for a smart and interactive environment: a survey and future directions. Sensors (Basel, Switzerland) 19(19):4330. https://doi.org/10.3390/s19194330
5. Grubert Jens, Langlotz Tobias, Zollmann Stefanie, Regenbrecht Holger (2016) Towards pervasive augmented reality: context-awareness in augmented reality. IEEE Trans Visual Comput Graph 23:1

Printed in the United States
By Bookmasters